VOLATILE ORGANIC COMPOUND ANALYSIS IN BIOMEDICAL DIAGNOSIS APPLICATIONS

VOLATILE ORGANIC COMPOUND ANALYSIS IN BIOMEDICAL DIAGNOSIS APPLICATIONS

Edited by
Raquel Cumeras, PhD
Xavier Correig, PhD

Apple Academic Press Inc.   Apple Academic Press Inc.
3333 Mistwell Crescent   9 Spinnaker Way
Oakville, ON L6L 0A2 Canada   Waretown, NJ 08758 USA

© 2019 by Apple Academic Press, Inc.

First issued in paperback 2021

Exclusive worldwide distribution by CRC Press, a member of Taylor & Francis Group
No claim to original U.S. Government works

ISBN 13: 978-1-77463-426-4 (pbk)
ISBN 13: 978-1-77188-744-1 (hbk)

Library and Archives Canada Cataloguing in Publication

Volatile organic compound analysis in biomedical diagnosis applications / edited by Raquel Cumeras, PhD, Xavier Correig, PhD.

Includes bibliographical references and index.
Issued in print and electronic formats.
ISBN 978-1-77188-744-1 (hardcover).--ISBN 978-0-429-43358-0 (PDF)

1. Volatile organic compounds--Diagnostic use. 2. Metabolites. 3. Diagnosis. 4. Systems biology. I. Cumeras, Raquel, editor II. Correig, Xavier, editor

| QP171.V65 2018 | 616.07'5 | C2018-904383-0 | C2018-904384-9 |

Library of Congress Cataloging-in-Publication Data

Names: Cumeras, Raquel, editor. | Correig, Xavier, editor.

Title: Volatile organic compound analysis in biomedical diagnosis applications / editors, Raquel Cumeras, Xavier Correig.

Description: Toronto ; New Jersey : Apple Academic Press, 2019. | Includes bibliographical references and index.

Identifiers: LCCN 2018034925 (print) | LCCN 2018035426 (ebook) | ISBN 9780429433580 (ebook) | ISBN 9781771887441 (hardcover : alk. paper)

Subjects: | MESH: Metabolomics--methods | Volatile Organic Compounds--analysis | Diagnostic Techniques and Procedures | Volatile Organic Compounds--metabolism

Classification: LCC QP171 (ebook) | LCC QP171 (print) | NLM QU 120 | DDC 572/.4--dc23

LC record available at https://lccn.loc.gov/2018034925

Apple Academic Press also publishes its books in a variety of electronic formats. Some content that appears in print may not be available in electronic format. For information about Apple Academic Press products, visit our website at **www.appleacademicpress.com** and the CRC Press website at **www.crcpress.com**

ABOUT THE EDITORS

Raquel Cumeras, PhD, has received a starting grant from the Martí Franquès Programme at the University Rovira i Virgili (URV, Tarragona, Catalonia, Spain) in 2017 to identify reliable and specific urinary bladder cancer biomarkers using metabolomics and crosslinking them with other urinary cancers (prostate and kidney). She has a broad knowledge in pre-concentration methods and sensors as well as in analytical chemistry for volatile compounds, including standard techniques like GC-MS and gas sensors based in ion mobility spectrometry sensors. The high quality of her research has been widely proven. She has been awarded with a postdoctoral starting grant, her PhD thesis received the Outstanding PhD thesis award, and also, the results of her research have been published in high-standard peer-reviewed scientific journals. Recently, she has submitted a new paper on metabolomics analysis as first author and a review of VOCs. After completing her PhD, she spent two years as a postdoctoral scholar at the University of California Davis, and after a sabbatical year for maternity, she did a postdoc at the Institute of Bioengineering of Catalonia (IBEC, Barcelona, Catalonia, Spain). Nowadays, her own research is focused not only in the development of metabolomics studies for determining volatile biomarkers but also on cancer metabolomics studies using the less invasive samples as breath, urine, blood, and feces.

Xavier Correig, PhD, is a Full Professor at the Universitat Rovira i Virgili (URV, Tarragona, Catalonia, Spain) and has extensive experience in training at the Department of Electronic Engineering and Research. He created and coordinated the Metabolomics Platform (http://metabolomicsplatform.com/), a research infrastructure created by the URV and the CIBERDEM. With more than 10 years of experience in the field, he is an expert on metabolomics, especially in NMR spectroscopy and mass spectrometry imaging. Professor Correig has published more than 200 articles, supervised several PhD students, and created a spin-off company (Biosfer Teslab), which is commercializing the Liposcale Test (a novel advanced lipoprotein test based on 2D diffusion-ordered 1H NMR spectroscopy). Also, he has an experience of more than 20 years in gas sensing and electronic nose based on metal oxides sensors or mass spectrometry.

CONTENTS

List of Contributors... *ix*

List of Abbreviations .... *xiii*

Acknowledgments..*xvii*

Preface ....*xix*

PART I: Volatiles and Detection Technologies **1**

1. **The Volatilome in Metabolomics** **3**
    Raquel Cumeras and Xavier Correig

2. **Basics of Gas Chromatography Mass Spectrometry System**.............. **31**
    William Hon Kit Cheung and Raquel Cumeras

PART II: Biomedical Diagnosis Applications................................ **51**

3. **Analysis of Volatile Organic Compounds for Cancer Diagnosis**............. **53**
    Abigail V. Rutter and Josep Sulé-Suso

4. **Artificial Olfactory Systems Can Detect Unique Odorant Signature of Cancerous Volatile Compounds** **79**
    Radu Ionescu

5. **Bottom-Up Cell Culture Models to Elucidate Human *In Vitro* Biomarkers of Infection**.. **105**
    Michael Schivo, Mitchell M. McCartney, Mei S. Yamaguchi, Eva Borras, and Cristina E. Davis

6. **Characterizing Outdoor Air Using Microbial Volatile Organic Compounds (MVOCs)** **123**
    Sonia Garcia-Alcega and Frédéric Coulon

7. **Breathomics and Its Application for Disease Diagnosis: A Review of Analytical Techniques and Approaches** **159**
    David J. Beale, Oliver A. H. Jones, Avinash V. Karpe, Ding Y. Oh, Iain R. White, Konstantinos A. Kouremenos, and Enzo A. Palombo

PART III: Computational Tools **195**

8. **The Need of External Validation for Metabolomics Predictive Models** **197**

    Raquel Rodríguez-Pérez, Marta Padilla, and Santiago Marco

9. **Insight into KNApSAcK Metabolite Ecology Database: A Comprehensive Source of Species: VOC-Biological Activity Relationships** **225**

    Azian Azamimi Abdullah, Md. Altaf-Ul-Amin, and Shigehiko Kanaya

Index *249*

LIST OF CONTRIBUTORS

Md. Altaf-Ul-Amin
Graduate School of Information Science, Nara Institute of Science and Technology (NAIST),
8916–5, Takayama, Ikoma, 630–0192 Nara, Japan

Azian Azamimi Abdullah
Graduate School of Information Science, Nara Institute of Science and Technology (NAIST),
8916–5, Takayama, Ikoma, 630–0192 Nara, Japan
School of Mechatronic Engineering, Universiti Malaysia Perlis (UniMAP),
Pauh Putra Campus, 02600 Arau, Perlis, Malaysia

David J. Beale
Commonwealth Scientific and Industrial Research Organization (CSIRO), Land & Water,
P.O. Box 2583, Brisbane, Queensland 4001, Australia. Email: david.beale@csiro.au (D.J.B.).
Phone: +61 7 3833 5774

Eva Borras
Department of Mechanical and Aerospace Engineering, University of California, Davis, One Shields
Avenue, Davis, CA, 95616 USA.

William Hon Kit Cheung
Department of Appied science, Northumbria Univeristy, Newcastle upon Tyne, NE1 8ST, UK,
William.Cheung@northumbria.ac.uk, +447468580538

Xavier Correig
Metabolomics Platform, Metabolomics Interdisciplinary Laboratory (MiL@b),
Department of Electronic Engineering (DEEEA), Universitat Rovira i Virgili, IISPV,
43003 Tarragona, Catalonia, Spain, Tel.: +34.977559623, E-mail: xavier.correig@urv.cat

Frédéric Coulon
Cranfield University, School of Water, Energy and Environment, Cranfield, MK43 0AL, UK,
E-mail: f.coulon@cranfield.ac.uk. Tel: +44 (0) 1234 75 4981

Raquel Cumeras
Metabolomics Platform, Metabolomics Interdisciplinary Laboratory (MiL@b), Department of
Electronic Engineering (DEEEA), Universitat Rovira i Virgili, IISPV, 43003 Tarragona, Catalonia,
Spain, Tel.: +34.977558764, E-mail: raquel.cumeras@urv.cat
Biomedical Research Centre in Diabetes and Associated Metabolic Disorders (CIBERDEM), 28029
Madrid, Spain
Signal and Information Processing for Sensing Systems, Institute for Bioengineering of Catalonia
(IBEC), Baldiri Reixac 4–8, 08028-Barcelona, Spain

Cristina E. Davis
Department of Mechanical and Aerospace Engineering, University of California, Davis, One Shields
Avenue, Davis, CA, 95616 USA, E-mail: cedavis@ucdavis.edu

Sonia Garcia-Alcega
Cranfield University, School of Water, Energy and Environment, Cranfield, MK43 0AL, UK

Radu Ionescu
Universitat Rovira i Virgili, Department of Electronic Engineering, Av. Països Catalans 26, 43007
Tarragona, Spain, Tel.: +34–977–55–87–54, Fax: +34–999–55–96–05, E-mail. radu.ionescu@urv.cat

Oliver A. H. Jones
Australian Centre for Research on Separation Science, School of Science, RMIT University,
PO Box 2547, Melbourne, Victoria 3000, Australia, E-mail: oliver.jones@rmit.edu.au

Shigehiko Kanaya
Graduate School of Information Science, Nara Institute of Science and Technology (NAIST),
8916–5, Takayama, Ikoma, 630–0192 Nara, Japan, Tel.: +81–743–72–5952, Fax: +81–743–72–5329,
E-mail: skanaya@gtc.naist.jp

Avinash V. Karpe
Department of Chemistry and Biotechnology, Swinburne University of Technology. PO Box 218,
Hawthorn, Victoria 3122, Australia, E-mail: akarpe@swin.edu.au

Konstantinos A. Kouremenos
Metabolomics Australia, Bio21 Molecular Science and Biotechnology Institute,
University of Melbourne, 30 Flemington Road, Parkville, VIC, 3010, Australia,
E-mail: konstantinos.kouremenos@unimelb.edu.au

Mitchell M. McCartney
Department of Mechanical and Aerospace Engineering, University of California, Davis, One Shields
Avenue, Davis, CA, 95616 USA

Santiago Marco
Signal and Information Processing for Sensing Systems, Institute for Bioengineering of Catalonia,
Baldiri Reixac 4–8, 08028-Barcelona, Spain
Department of Engineering: Electronics, Universitat de Barcelona,
Martí i Franqués 1, 08028-Barcelona, Spain, E-mail: smarco@ibecbarcelona.eu

Ding Y. Oh
WHO Collaborating Centre for Reference and Research on Influenza (VIDRL), Peter Doherty
Institute for Infection and Immunity, 792 Elizabeth Street, Melbourne, VIC, Australia, and School of
Applied and Biomedical Sciences, Federation University, Churchill, Victoria, Australia,
E-mail: dingthomas.oh@influenzacentre.org

Marta Padilla
Signal and Information Processing for Sensing Systems, Institute for Bioengineering of Catalonia,
Baldiri Reixac 4–8, 08028-Barcelona, Spain

Enzo A. Palombo
Department of Chemistry and Biotechnology, Swinburne University of Technology. PO Box 218,
Hawthorn, Victoria 3122, Australia, E-mail: epalombo@swin.edu.au

Raquel Rodríguez-Pérez
Signal and Information Processing for Sensing Systems, Institute for Bioengineering of Catalonia,
Baldiri Reixac 4–8, 08028-Barcelona, Spain
Department of Engineering: Electronics, Universitat de Barcelona,
Martí i Franqués 1, 08028-Barcelona, Spain

Abigail V. Rutter
Institute for Science and Technology in Medicine, Keele University, Guy Hilton Research Centre,
Thornburrow Drive, Stoke on Trent ST4 7QB, UK

Michael Schivo
Department of Medicine, Division of Pulmonary, Critical Care, and Sleep Medicine, University of
California, Davis, Sacramento, CA, USA.
Center for Comparative Respiratory Biology and Medicine, University of California, Davis, Davis,
CA, USA

Josep Sulé-Suso
Institute for Science and Technology in Medicine, Keele University, Guy Hilton Research Centre,
Thornburrow Drive, Stoke on Trent ST4 7QB, UK
Oncology Department, Royal Stoke University Hospital, University Hospitals of North Midlands
(UHNM), Newcastle Rd, Stoke on Trent, Staffordshire ST4 6QG, UK,
E-mail: Josep.SuleSuso@uhnm.nhs.uk

Iain R. White
Manchester Institute of Biotechnology, University of Manchester, Princess St, Manchester M1 7DN,
United Kingdom, E-mail: iain.white-2@manchester.ac.uk

Mei S. Yamaguchi
Department of Mechanical and Aerospace Engineering, University of California, Davis,
One Shields Avenue, Davis, CA, 95616, USA

LIST OF ABBREVIATIONS

AOSs	artificial olfaction systems
ANOVA	analysis of variance
AMDIS	automated mass spectral deconvolution and identification system
ARDS	acute respiratory distress syndrome
ATP	adenosine triphosphate
AUC	area under the curve ROC curve
BC	breast cancer
BCA	breath collection apparatus
BIONOTE	biosensor-based multisensorial system for mimicking nose
BMAS	bioaerosol mass spectrometer
BP	boiling point
C_ID	compound ID
CC	colorectal cancer
COPD	chronic obstructive pulmonary disease
CSIRO	Commonwealth Scientific and Industrial Research Organization
CTC	circulating tumor cells
DART	direct analysis in real time
DBs	databases
DC	direct current
DIMS	direct injection mass spectrometry
EI	electron impact ionization
ENs	electronic noses
ENO	exhaled nitric oxide
CAS	chemical abstracts service
CAS RN	CAS registry number
CC	compensated cirrhosis
CE	capillary electrophoresis
CI	chemical ionization
CID	collision induced disassociation

CKD	chronic kidney disease
CLD	chronic liver disease
CODH	carbon monoxide
CODH	carbon monoxide dehydrogenase
COPD	chronic obstructive pulmonary disease
CPAP	continuous positive airways pressure
CT	computed tomography
CV	coefficients of variability
CV	cross-validation
EBC	exhaled breath condensate
EBV	exhaled breath vapor
ECG	electrocardiogram
EDRN	early detection research network
ELISA	enzyme-linked immunosorbent assay
ES	effect size
FAIMS	high field asymmetric ion mobility spectrometry
FDR	false discovery rate
FET	field effect transistors
FT-ICR	Fourier transforms ion cyclotrons resonance
FT-IR	Fourier transform infrared spectroscopy
FWER	family-wise error rate
GC	gas chromatography
GC-MS	gas chromatography mass spectrometry
GI	gastrointestinal
HAPE	high altitude pulmonary edema
HASM	human smooth muscle
HCA	hierarchical cluster analysis
HETP	height equivalent to theatrical plates
HMDB	human metabolome database
ICP-MS	inductively coupled plasma-mass spectrometry
ID	identification
IPAH	idiopathic pulmonary arterial hypertension
IR	infra-red
IUPAC	international union of pure and applied chemistry
k-NN	k-nearest neighbors
KEGG	Kyoto Encyclopedia of Genes and Genomes
LDA	linear discriminant analysis
LC	liquid chromatography

LC	lung cancer
LC-MS	liquid chromatography-mass spectrometry
LDA	linear discriminant analysis
LFDR	local false discovery rate
LOD	limit of detection
LTMs	low thermal mass system
LV	latent variables
MD	mahalanobis distances
MS	mass spectrometry
MSI	metabolomics standards initiative
MsPGN	mesangial proliferative glomerulonephritis
MSW	municipal solid waste treatment
MVOC	microbial volatile organic compound
mVOC	MVOC database
NA-NOSE	nanoparticles sensors-based e-NOSE
NHDF	normal human diploid fibroblasts
NIR	near-infrared spectroscopy
NMR	nuclear magnetic resonance
NO	nitric oxide
NSCLC	non-small-cell lung carcinoma
OLGIM	operative link on gastric intestinal metaplasia
OPLS-DA	orthogonal projections to latent structures discriminant analysis
PAH	pulmonary arterial hypertension
PC	prostate cancer
PCA	principal component analysis
PLFAs	phospholipid fatty acids
PLS	partial least-squares
PLS-DA	partial least-squares discriminant analysis
PM	particulate organic matter
ppb	parts-per-billion
ppt	parts-per-trillion
PTR-MS	proton-transfer-reaction mass spectrometry
QC	quality control
QCM	quartz crystal microbalance
QMF	quadrupole mass filters
QQQ	triple quadrupole mass filters
qTOF	quadrupole time of flight

RECIST	response evaluation criteria in solid tumors
ReCIVA	respiration collector for in vitro analysis
ROS	reactive oxygen species
RT	retention time
SAW	surface acoustic wave
SELDI-TOF-MS	surface-enhanced laser desorption/ionization TOF-MS
SESI-MS	secondary electrospray ionization-mass spectrometry
SIBS	spectral intensity bioaerosol sensor
SIFT-MS	selected ion flow tube-mass spectrometry
SIM	selective ion monitoring
SOM	self-organization mapping
SPME	solid phase micro extraction
SRM	selective reaction monitoring
SVOC	semi-volatile organic compounds
SVM	support vector machine
TCA	tricarboxylic acid
TD	thermal desorption tubes
TOF	time of flight
UV	ultraviolet
UBT	urea breath test
VOC	volatile organic compound
VOCC	volatile organic compound cancer database
VVOC	very volatile organic compounds
WHO	World Health Organization
WIBS	wideband integrated bioaerosol sensor
WWTP	wastewater treatment plants

ACKNOWLEDGMENTS

The editors express their sincere gratitude to the chapter authors who contributed their valuable time and expertise to this book. Without their support, this book would not have become reality.

Dr. Cumeras acknowledges the project 2017PMF-POST-10 from the Martí-Franquès Programme (URV, Spain); and Prof. Correig acknowledges the project TEC2015–69076-P financed by the Ministerio de Economía y Competitividad (Spain).

PREFACE

Volatile organic compounds (VOCs) are gaining considerable interest, and research is being done to obtain insight into underlying mechanisms of physiological and pathophysiological processes and to exploit concentration profiles of VOCs in biological sources for disease detection and/or therapeutic monitoring of treatment regimens. As a biochemical probe, VOCs are unique as they can provide both noninvasive and continuous information on the metabolic/physiological state of an individual. In humans, VOCs are found in exhaled breath and in a broad variety of biofluids, like urine, blood, tissue, skin emanations, saliva, and feces. Detection and confirmation of VOCs as biomarkers may be additionally complicated by the fact that VOC levels are linked to metabolic processes. Therefore, metabolomics of VOCs (known also as volatilomics) has gained interest and currently is a hot topic in the metabolomics field.

Metabolomics can be defined as a new emerging -omic science in systems biology that is aimed to decipher the metabolic profile in complex systems through the combination of data-rich analytical techniques (NMR, MS) and multivariate data analysis. Techniques used include liquid chromatography (LC-MS) and nuclear magnetic resonance (NMR) for the liquid fraction, and for volatiles with gas chromatography (GC-MS), proton transfer reaction (PTR/MS), selected ion flow tube (SIFT)/MS, ion mobility spectrometry (IMS), and electronic noses.

The purpose of this book is to provide a state-of-the-art picture of how volatile compounds are used in metabolomics, to give authoritative education and guidance on the analytical and statistical techniques used, and to identify and review the main current areas of application. It should be of interest to analytical scientists and to those working in application areas as diverse as plant science, environmental science, and human clinical medicine. Volatiles have the advantage of being relieved from noninvasive samples like breath and urine and are of great interest in point-of-care medicine researchers. In summary, we believe that this volume should prove educational, informative, critical, and thought-provoking.

The book is composed of nine chapters and divided into three parts. Part one consists of two chapters. Chapter 1 provides a thorough state-of-art of VOCs in the biomedical area with metabolomics techniques and describes its potential in systems biology and precision medicine, and also gives a standard reporting metadata list for the data recorded in volatile metabolomics experiments. And Chapter 2 provides a thorough background of the gold standard technique for volatiles analysis, the GC-MS.

Part two consists of five chapters providing some applications in which volatiles are used. Chapter 3 describes the uses of VOCs in breath for different cancers and how they can be clinically implemented. Chapter 4 describes the use of artificial olfactory systems for cancer-related VOCs. Chapter 5 describes the microbial VOCs specifying the respiratory infections. Chapter 6 describes the use of microbial VOCs to chemically characterize ambient bioaerosols and identify pathogens in air; and Chapter 7 describes all the instrumentation and analytical techniques used for the analysis of different illnesses with breath samples.

Part three consists of two chapters. Chapter 8 describes the need of external validation for metabolomics predictive models; and finally Chapter 9 describes the KNApSAcK metabolite ecology database of VOCs emitted by various living organisms including microorganisms, fungi, plants, animals, and humans.

—**Raquel Cumeras, PhD**
Xavier Correig, PhD

PART I
Volatiles and Detection Technologies

CHAPTER 1

THE VOLATILOME IN METABOLOMICS

RAQUEL CUMERAS[1,3,4] and XAVIER CORREIG[2,3]

[1]Metabolomics Platform, Metabolomics Interdisciplinary Laboratory (MiL@b), Department of Electronic Engineering (DEEEA), Universitat Rovira i Virgili, IISPV, 43003 Tarragona, Catalonia, Spain, Tel.: +34.977558764, E-mail: raquel.cumeras@urv.cat

[2]Metabolomics Platform, Metabolomics Interdisciplinary Laboratory (MiL@b), Department of Electronic Engineering (DEEEA), Universitat Rovira i Virgili, IISPV, 43003 Tarragona, Catalonia, Spain, Tel.: +34.977559623, E-mail: xavier.correig@urv.cat

[3]Biomedical Research Centre in Diabetes and Associated Metabolic Disorders (CIBERDEM), 28029 Madrid, Spain

[4]Signal and Information Processing for Sensing Systems, Institute for Bioengineering of Catalonia (IBEC), Baldiri Reixac 4–8, 08028-Barcelona, Spain

ABSTRACT

This chapter is intended to provide a broad vision of what are volatile organic compounds (VOCs), the reasons of their interest in biological matrices, and their use in the study of the VOCs produced by all living organisms, the volatilomics. We provide a list of the last review for the period 2016–2017, focusing on biomedical applications as well as sampling techniques and state-of-the-art scientific instrumentation used for those applications. We specially focus on human volatilomics and its use with metabolomics techniques, and their relation with systems biology and precision medicine. Finally, as volatilomics generates large and diverse sets of analytical data, it has significant challenges, mainly in standardization and metadata reporting, which are described in detail.

1.1 INTRODUCTION

1.1.1 VOLATILE ORGANIC COMPOUNDS

What are volatile organic compounds (VOCs)? Depending on the application sector or country, a different definition exists. Traditionally, VOCs have been associated with indoor air pollutants adversely impacting people's health as with the formation of ground level ozone and particulate matter, the main ingredients of smog contamination. Consequently, environmental and atmospheric VOCs have received thorough attention, as several regulations exist (EU CE, 2010; US EPA, 2014). However, a group of World Health Organization experts (WHO, 1989) provided an internationally recognized definition of VOCs. It was specified that VOCs concept could be classified for their boiling points (BP). The lower the BP, the higher the volatility, more likely the compound will be emitted. Therefore, VOCs have a BP range from 50–100°C to 240–260°C, sampled by adsorption and analyzed by gas chromatography (GC) or liquid chromatography (LC). Outside this interval, lower BP is named very volatile organic compounds (VVOC, <0°C to 50–100°C) but those compounds have such low BP that they are difficult to measure and are found almost entirely as gases. The compounds with higher BP are named semi-volatile organic compounds (SVOC, 240–260°C to 380°C), and finally, compounds with BP > 380°C, are named particulate organic matter (PM).

Also, VOCs can be classified from their origin.

1. *Anthropogenic VOCs* (xenobiotic, man-made) that can be found in paints, adhesives, petroleum products, pharmaceuticals, refrigerants, as compounds of fuels, solvents, etc. Some of those VOCs are hazardous to human health or can damage the environment. For this reason, they are regulated by law, specifically indoors (where concentrations are the highest) because long-term health effects exist even that harmful VOCs typically are not acutely toxic.
2. *Biogenic VOCs* (BVOC) are organic atmospheric trace gases other than carbon dioxide and monoxide (Kesselmeier and Staudt, 1999).

1.1.2   THE VOLATILOME

Volatilome or volatolome is the totality of VOCs produced by all living organisms (plants, animals, and so forth), as result products of bio functions happening in the living organisms, including cells, organs, microbiome, etc. The volatilome compounds can be classified from their origin: exogenous (due to external conditions); and endogenous (due to normal bio-function).

Living organisms produce exogenous VOCs metabolites as a result of the interaction with an external exposition. Exogenous compounds enter the organism via inhalation, ingestion or dermally, and then contribute to the produced metabolites. Exogenous sources encompass the environment (including pollutants) (Filipiak et al., 2012), nutrition (Gibney et al., 2005), drugs (Cumeras et al., 2013), smoking habit (Filipiak et al., 2012), and even ionizing radiation (Menon et al., 2016), among others.

Living organisms naturally produce endogenous VOCs metabolites. However, finding the same metabolite in different body fluids does not mean that they are of endogenous origin (Schmidt and Podmore, 2015). Some inhaled VOCs may bind/dissolve in hemoglobin (Poulin and Krishnan, 1996) and be stored in body compartments and later exhaled/excreted through urine (Silva et al., 2011). VOCs plants emissions play an important role in plant–plant interactions (Baldwin et al., 2006), plant–microbial interactions (Junker and Tholl, 2013), and herbivores-induced VOCs (Heil, 2014; Kessler and Baldwin, 2001). Cells (Filipiak et al., 2016), microbial (Bos et al., 2013; Cumeras et al., 2016; Lemfack et al., 2014; Li et al., 2016), and microbial interactions (Schmidt et al., 2015, 2016; Tyc et al., 2015) are emitting VOCs depending on the growing stage or the medium they grew (Heddergott et al., 2014). Animals have shown specific VOCs emission, as recently reported for dolphins (Aksenov et al., 2014b) and whales (Cumeras et al., 2014).

The volatilome can be analyzed by a number of techniques ranging in sophistication and performance: from reference techniques based on chromatographic separation and mass spectrometry (GC/MS or LC/MS) regularly used in metabolomics, to less complex, instrumentation based in *ion mobility spectrometry* (IMS and high field asymmetric ion mobility spectrometry (FAIMS)) and eventually *chemical sensor arrays* as in e-noses, for maximum integration and minimum cost. Metabolomics studies the small molecule metabolic products (the metabolome) of a biological system (cell, tissue, organ, biological fluid, or organism) at a specific point in time.

1.2   CURRENT REVIEWS

Another noteworthy recent development is the appearance of review articles on volatilome, as summarized in Table 1.1 for the human volatilome and in Table 1.2 for sampling/detection techniques including metabolomics procedures. The large number of reviews published in the last two years shows how volatile metabolomics is on its height (or ascending) momentum. Upto 34 reviews have been found in the literature reporting any aspect affecting the biomedical volatilome (out of which, 21 reviews are about human healthy and disease VOCs, and 13 reviews are about sampling and detection techniques).

TABLE 1.1   Review Articles on Human Volatilome, for the Period 2016–2017 (Including Those Published in Early 2018, But Available Online in 2017)

Title of Review Article	Reference
A compendium of volatile organic compounds (VOCs) released by human cell lines.	Filipiak et al. (2016)
A review of analytical techniques and their application in disease diagnosis in breathomics and salivaomics research.	Beale et al. (2017)
Breathomics from exhaled volatile organic compounds in pediatric asthma.	Neerincx et al. (2017)
Detection of halitosis in breath: Between the past, present, and future.	Nakhleh et al. (2017)
Detection of volatile organic compounds (VOCs) from exhaled breath as noninvasive methods for cancer diagnosis.	Sun et al. (2016)
Electronic noses for well-being: Breath analysis and energy expenditure.	Gardner and Vincent (2016)
Electronic nose technology in respiratory diseases.	Dragonieri et al. (2017)
Exhaled breath analysis: a review of 'breath-taking' methods for off-line analysis.	Lawal et al. (2017)
GC–MS based metabolomics used for the identification of cancer volatile organic compounds as biomarkers.	Lubes and Goodarzi (2018)
Ion mobility in clinical analysis: Current progress and future perspectives.	Chouinard et al. (2016)

TABLE 1.1 *(Continued)*

Title of Review Article	Reference
In vitro assays as a tool for determination of VOCs toxic effect on respiratory system: A critical review.	Gałęzowska et al. (2016)
Metabolomics: a state-of-the-art technology for better understanding of male infertility.	Minai-Tehrani et al. (2016)
Review of recent developments in determining volatile organic compounds in exhaled breath as biomarkers for lung cancer diagnosis.	Zhou et al. (2017)
Solid-phase microextraction technology for *in vitro* and *in vivo* metabolite analysis.	Zhang et al. (2016)
Systematic approaches for biodiagnostics using exhaled air.	Shende et al. (2017)
The why and how of amino acid analytics in cancer diagnostics and therapy.	Manig et al. (2017)
Translation of exhaled breath volatile analyses to sport and exercise applications.	Heaney and Lindley (2017)
VOC breathe biomarkers in lung cancer.	Saalberg and Wolff (2016)
Volatile organic compounds as new biomarkers for colorectal cancer: a review.	Di Lena et al. (2016)
Volatile organic compound detection as a potential means of diagnosing cutaneous wound infections.	Ashrafi et al. (2017)
Volatile organic compounds in asthma diagnosis: a systematic review and meta-analysis.	Cavaleiro Rufo et al. (2016)

TABLE 1.2 Review Articles on Volatilomics Sampling, Detection Technologies or Metabolomics Procedures, for the Period 2016–2017 (Including Those Published in Early 2018, But Available Online in 2017)

Title of Review Article	Reference
A critical review on the diverse preconcentration procedures on bag samples in the quantitation of volatile organic compounds from cigarette smoke and other combustion samples.	Kim et al. (2016)
Advances in electronic-nose technologies for the detection of volatile biomarker metabolites in the human breath.	Wilson (2017)
Artificial Nose Technology: Status and Prospects in Diagnostics.	Fitzgerald et al. (2017)
Biomarker metabolite signatures pave the way for electronic-nose applications in early clinical disease diagnoses.	Alphus Dan (2016)
Detection of hazardous volatile organic compounds (VOCs) by metal oxide nanostructures-based gas sensors: A review.	Mirzaei et al. (2016)

TABLE 1.2   *(Continued)*

Title of Review Article	Reference
Extraction media used in needle trap devices—Progress in development and application.	Kędziora and Wasiak (2017)
Metabolomics for the masses: The future of metabolomics in a personalized world.	
Navigating freely-available software tools for metabolomics analysis.	Spicer et al. (2017b)
Quality assurance procedures for mass spectrometry untargeted metabolomics: A review.	Dudzik et al. (2018)
Recent advances in engineered graphene and composites for detection of volatile organic compounds (VOCs) and non-invasive diseases diagnosis.	Tripathi et al. (2016)
Recent advances in liquid and gas chromatography methodology for extending coverage of the metabolome.	Haggarty and Burgess (2017)
Recent applications of gas chromatography with high-resolution mass spectrometry.	Spánik and Machyňáková (2018)
The analytical process to search for metabolomics biomarkers.	Luque de Castro and Priego-Capote (2018)

1.3 THE HUMAN VOLATILOME

The human volatilome encompasses the totality of VOCs produced by human beings. The origin and metabolic outcome of the completely emitted VOCs have not yet been properly elucidated in sufficient depth (Haick et al., 2014). The human volatilome VOCs can be classified depending on their etiology (Cumeras, 2017) into three categories:

- VOC's from healthy individuals (endogenous);
- induced VOC's (exogenous); and
- VOC's from diseased individuals (endogenous/exogenous).

1.3.1 HEALTHY (ENDOGENOUS) HUMAN VOCs

Healthy volatiles are important as *knowing what is normal will help to assess what is not* (Cumeras, 2017). De Lacy Costello (2014) reviewed upto 1840 human VOCs:

- blood (serum and plasma) – 154;
- breath – 872;
- feces – 381;
- milk – 256;
- saliva – 359;
- skin secretions (sweat and follicle fluids) – 532; and
- urine – 279.

This list has to be understood as alive, as more compounds will be added or removed with new studies and by scanning of other biofluids at the same level than breath experiments. In the future, this list will include scanning the volatile content of other biofluids like cerumen (Prokop-Prigge et al., 2014), vaginal fluid (Michael et al., 1974), tears (Muñoz-Hernández et al., 2016), pleural fluid (Huang et al., 2016), etc.

Also, the amount of VOCs detected will depend on the sampling (Berkhout et al., 2016), their transport and storage conditions, and/or the detection techniques used. So, procedures standardization is the key. Multi-based cohort studios that take into account of geographical origin (Haitham et al., 2013) and gender (Das et al., 2014) differences should be done, to establish that the detected VOC profile is representative of a healthy individual condition.

1.3.2   INDUCED (EXOGENOUS) HUMAN VOLATILES

Induced (exogenous) human volatiles are due to non-harmful external conditions. Those will include the resulting VOCs from the exposure to:

- drugs (Cumeras et al., 2013) – studied in pharmacometabolomics;
- environment (excluding pollutants) (Haitham et al., 2013);
- exposure to ionizing radiation (Menon et al., 2016) – studied in radiation metabolomics;
- microbials (excluding pathogens) (Petrof et al., 2013) – studied in microbiome; and
- nutrition (Gibney et al., 2005) – studied in foodomics.

Establishing cause–effect relationship on exogenous human volatiles would not always be possible. Those induced (exogenous) human volatiles by their own will not produce a health effect, if the concentration and exposition are low. However, if this exposition is prolonged over the time, by over-eating, an accumulative load of stressors in the environment, other excesses, psychological and emotional factors (over-work, anger, fear, sadness, etc.) may lead to fatigue and/or epigenetic expressions. We are stressing out the immune system and may lead to disease. Also, the harmful exogenous compounds are included in the disease category.

The study of the induced (exogenous) human volatiles is known as exposome, and it was first defined in 2005 by Wild as "exposome encompasses life-course environmental exposures (including lifestyle factors), from the prenatal period onwards" (Wild, 2005). However, the most accepted one is that "exposome is the cumulative measure of environmental influences and associated biological responses throughout the lifespan, including exposures from the environment, diet, behavior, and endogenous processes" (Miller and Jones, 2014).

1.3.3 DISEASE (ENDOGENOUS/EXOGENOUS ORIGIN) ASSOCIATED HUMAN VOLATILES

As the metabolites associated to a disease or metabolic disorder can help to improve its diagnosis, prognosis or the stratification of patients, they are the most studied human volatiles (Sornette et al., 2009). The list of diseases or metabolic disorders that have been analyzed with volatilomics is resumed in Table 1.3. This is not an exhaustive list, and more VOCs related to specific diseases or disorders are explored every day.

TABLE 1.3 Diseases and Metabolic Disorders That Have Been Studied with Volatilomics

Type	Reference
Bacteremia (bacteria presence in the blood)	Christiaan et al. (2016)
Digestive system	
- Celiac disease	Francavilla et al. (2014)
- Colorectal cancer	de Boer et al. (2014; Di Lena et al. (2016)
- Crohn disease	Cauchi et al. (2014)
- Endoluminal gastrointestinal diseases	Markar et al. (2015)
- Gastric cancer	Daniel and Thangavel (2016)
- Inflammatory bowel diseases	Kurada et al. (2015)
- Liver cancer	Xue et al. (2008)
- Oral cancer	Bouza et al. (2017)

TABLE 1.3   *(Continued)*

Type	Reference
- Pancreatic cancer	Navaneethan et al. (2014)
- Ulcerative colitis	Cauchi et al. (2014)
Gender specific	
- ♀ Breat cancer	Michael et al. (2010)
- ♀ Cervical cancer	Guerrero-Flores et al. (2017)
- ♂ Prostate cancer	Lima et al. (2016)
Glands	
- Thyroid cancer	Guo et al. (2015)
Metabolic disorders	
- Diabetes	Wenwen et al. (2015)
Multiple organs affected	
- Cystic fibrosis (CF)	Robroeks et al. (2010)
- Invasive aspergillosis (IA) (fungal)	Koo et al. (2014)
- Malaria (parasite)	Berna et al. (2015)
- Tuberculosis (bacteria)	Banday et al. (2011)
Neurological	
- Autism spectrum disorder	De Angelis et al. (2015)
- Neurodegenerative diseases	Ko et al. (2014)
Respiratory	
- Asthma	Cavaleiro Rufo et al. (2016); Neerincx et al. (2017); Pijnenburg and Szefler (2015)
- Chronic obstructive pulmonary disease (COPD)	Phillips et al. (2014)
- Halitosis	Nakhleh et al. (2017)
- Influenza (virus)	Aksenov et al. (2014a)
- Lung cancer	Horváth et al. (2009)
- Respiratory tract infections	Nizio et al. (2016)
- Rhinovirus (virus)	Schivo et al. (2014)
Skin	
- Melanoma	Abaffy et al. (2013)
- Ulcers	Schivo et al. (2017)
Squamous cell carcinoma of the head and neck	Gruber et al., 2014)
Urinary	
- Bladder cancer	Jobu et al. (2012)
- Kidney mesangial proliferative glomerulonephritis (MsPGN)	Wang et al. (2015)

A disease can be triggered by being a result of an exogenous source (like products produced by the host in response to microbial infections), but also as a result of endogenous products of physiological/metabolic body processes already present in the body (like an urinary tract infection by residential bacterium *Escherichia coli* commonly found in the gastrointestinal (GI) tract). Many diseases are caused by microorganisms (bacteria's, fungi, parasites, viruses), and those microorganisms grow at the expense of the host, rather than diseases being spontaneously generated. Also, exogenous sources that produce or interfere with the diseases are being explored (mainly in breath analysis), for hospital related devices (like in ventilator-associated pneumonia) (Schnabel et al., 2015), smoking habit (Filipiak et al., 2012), and environment (pollutants) (Schnabel et al., 2015). However, not all published studies does an endogenous/exogenous assessment of the collected data or so-called metadata. If done, some other relevant factors like age, BMI, and gender might be found (Blanchet et al., 2017).

Additionally, it is not known which VOC's are produced or consumed by tumor cells as some of them may also be generated (or consumed) by non-cancerous cells (such as surrounding tissue cells or other regions of the body). Recently, Sousa et al. (2016) have unveiled an unknown cross-talk between tumor cells and surrounding cells to give them energy through autophagy by breaking down their own components into smaller metabolites, which are then released by the healthy cells so the tumor cells can consume them and have the energy to survive and grow. This cross-talk opens a new approach for cancer therapies, by stopping the nearby cells from releasing their metabolites, so the tumor cells stop growing.

1.4   FROM SYSTEMS BIOLOGY TO PRECISION MEDICINE

Systems biology (Ideker et al., 2001) examines the behavior and relationships of all of the components in a particular biological system while it is functioning. An alternative description (Bekri, 2016) is that "systems biology is the computational integration of data generated -omic platforms to understand function across different levels of biomolecular organization." The Human Genome Project unveiled the need to understand the relationship between genome by studying the products of the genome, namely proteins and expressed RNAs, such as tRNA and rRNA. This lead to the emergence of the transcriptomics and proteomics fields. This flowering of -omics sciences also spread to the low-molecular-weight compounds that carry out much of the cell's function like lipids, carbohydrates, vitamins, and hormones, among others. In parallel to the terms 'transcriptome' and proteome,' the set of metabolites synthesized by a biological system constitute its 'metabolome' (Fiehn, 2002). Volatilomics refers to the study of the volatile fraction of the metabolome. A reductionist view of every single -omic, will give an insight of what is happening in a precise level so that we can construct a whole model.

The ability of metabolomics to provide non-invasive translational biomarkers makes it an essential part of a systems biology approach. The capability to link changes in the metabolite profile to altered genes and/or proteins will help, in some cases, to elucidate the source of metabolite biomarkers. However, the relationship between health status, genetics, and metabolic state is highly complex and not easily determined. A systems biology approach provides a better understanding of diseases mechanisms and progression, and it can be used to identify early biomarkers of drug efficacy and toxicity (Tolstikov, 2016). The main outcome would be to find "the right drug for you" (Drew, 2016), the objective of precision medicine. From the AllofUS Research Program (US) (NIH, 2014), precision medicine is defined as "a groundbreaking approach to disease prevention and treatment based on people's individual differences in environment, genes, and lifestyle." A more inclusive definition would be that "precision medicine is the disease prevention and treatment based on people's individual differences in genome, proteome, transcriptome, metabolome, environment, and lifestyle."

As shown in Figure 1.1, systems biology and precision medicine are interdependent and based on the context of a patient's biological signatures (i.e., genetic, biochemical, etc.); clinicians would be able to select specific diagnostic testing and therapeutic intervention. This tailored approach to diagnosis and treatment has the potential to improve efficacy while limiting costs and side effects.

Spectrometry techniques are the gold standard tools for detecting quantitative VOC concentration profiles. However, for volatilomics diagnostics to enter clinical practice, approaches must be adopted that would allow implementing point-of-care devices, such as e-noses (Li et al., 2014), FAIMS (Covington et al., 2015), etc. Several bottlenecks slow-down the transition from conventional to precision medicine: generation of cost-effective high-throughput data; big data challenges (data access and management, data integration and interpretation, standardization, and validation); hybrid education and multidisciplinary teams; and individual and global economic relevance.

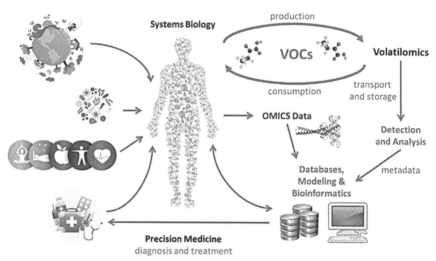

FIGURE 1.1   (See color insert.) Systems biology, volatilomics and precision medicine relations.

1.5   DATA ANALYSIS WORKFLOW AND STANDARDIZATION

With the availability of -omic technologies, researchers and clinicians are nowadays generating large amounts of data on biological samples. Big data bioinformatics technologies are now allowing for the analysis of those large data sets. Independently of the technique used (GC/MS, e-nose, FAIMS, etc.), the obtained data will need several steps for their analysis, as shown in Figure 1.2.

Metabolomics is amongst the most recent –omics so is lacking harmonization and protocol standardization (Bekri, 2016), and processes need to be thoroughly validated before implementing them for clinical and diagnostic purposes. In that line, the metabolomics standard initiative (MSI, http://www.metabolomics-msi.org/) and the coordination of standards in metabolomics (COSMOS, http://www.cosmos-fp7.eu/) aim to provide reporting standards of the biological system studied for all components of a metabolomics study including metadata as well as open standard encoded files and public repositories; however, compliance with those guidelines is low (Spicer et al., 2017a).

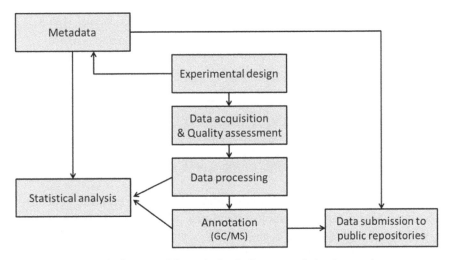

FIGURE 1.2   Block diagram of the typical volatilome metabolomic experiment.

1.5.1   DATA PROCESSING AND ANNOTATION

The different techniques used for voaltilome (GC/MS, FAIMS, e-nose, etc.) will produce specific data files. Volatile compounds are commonly analyzed with a GC-MS, a gas chromatograph coupled to a mass spectrometer. In fact, it is the gold standard technique as it provides chemical identification. In the following paragraphs, we describe the data processing for GC-MS data, once pre-processed as indicated in the following section.

1.5.1.1   MASS SPECTRAL LIBRARIES

Various libraries are used for the identification of the VOCs in the MS spectra (Table 1.4) (Milman and Zhurkovich, 2016), being the most used, the NIST and Wiley electronic collections. By combining different libraries, it will lead to correct identification results (Milman, 2015).

TABLE 1.4   Libraries/Data Sources for Volatile Analytes

Library/information source	Spectra	Compounds	Reference
Adams		2,205	Adams (2007)
Fiehn Lib	2212	>1,000	Kind et al. (2009)
Golm Metabolome Database		>2,000	Kopka et al. (2005)
MassBank	41,092	>15,000	Horai et al. (2010)
MassFinder		~2,000	Hochmuth (2012)
NIST 14 GC database	>385,872	>82,337	NIST (2014)
VocBinBase	1,537	3,435	Skogerson et al. (2011)
Wiley	>670,000	>3,000,000	Wiley (2014)

1.5.1.2   METABOLOMICS EXPERIMENTS DATABASE

Metabolomics experiments database and derived information (species, experimental platforms, metabolite standards, metabolite structures, protocols): MetaboLights (Haug et al., 2013) (www.ebi.ac.uk/metabolights/), Metabolomics Workbench (Sud et al., 2016) (www.metabolomicswork-bench.org/).

1.5.1.3   THE HUMAN METABOLOME DATABASE (HMDB)

The human metabolome database is a freely available electronic database containing detailed information about small molecule metabolites found in the human body. The database contains 42,003 metabolite entries including water-soluble and lipid soluble metabolites as well as metabolites that would be regarded as either abundant (>1 uM) or relatively rare (<1 nM). Additionally, 5,701 protein sequences are linked to these metabolite entries. Even though HMBD integrates databases of almost all the human tissues or biofluids, it doesn't have a specific entry for breath or volatilome.

1.5.1.4   HUMAN VOLATILOME DATABASES

A database of 1840 VOCs of healthy volatilome is available in the paper format (de Lacy Costello et al., 2014). A database of 551 cancer VOCs

(Agarwal et al., 2016) is available online at the VOCC, http://smagarwal. in/vocc/index.php. VOCC is a first of its kind attempt to provide a comprehensive non-redundant catalogue of VOCs involved in 17 cancer types. It thus provides a unique value-added resource in the field of cancer volatilomics.

Another useful volatilome databases, for its exogenous factor in humans, is the microbial volatile organic compound (mVOC) database (Lemfack et al., 2014) (http://bioinformatics.charite.de/mvoc/).

1.5.1.5   METABOLITES IDENTIFICATION

After the analysis with the MS, and checking the databases, 'possible' VOCs are ready to be classified. The formal definitions of metabolite annotation and identification, as developed by the Chemical Analysis Working Group of the Metabolomics Standards Initiative (MSI) (Sumner et al., 2007). The classification system defines 'identified compounds' (level 1), then 'putatively annotated compound' (level 2), 'putatively characterized compound classes' (level 3) and unknown (level 4, although unidentified or unclassified these metabolites can still be differentiated and quantified based upon spectral data).

1.5.2   METADATA

The minimum metadata for reporting biomedical volatile metabolomics will include:

I. Subjects

-   Ethical committee approval number (if applicable).
-   Location and/or hospital.
-   Medical history (disease or clinical symptoms; criteria for disease presence (e.g., surgical or pharmacological manipulation), medication, etc.).
-   Age range.
-   Weight and height range and/or BMI.
-   Gender.
-   Trial type (e.g., randomized trail, phase I–IV, etc.).

- Dietary restrictions and relevant control groups (if applicable).
- Further descriptors: genetic tests, ethnicity, smoking habit, blood pressure, anomalies in the habitual diet (e.g., vegetarian, habitual alcohol consumption), etc.

II. Experimental design

- A number of groups subjects/gender/group.
- Inclusion and exclusion criteria.
- Treatments (compound, route, dose, duration, vehicle) and/or fasting.
- Endpoints (subject outcome or sign that motivates the withdrawal of that individual from the studio, e.g., hematuria present in urine). Endpoint includes clinical chemistries, blood chemistry and hematology, and urine chemistry.

III. Sampling process and protocol

- Sample type (blood-serum (clotted)/plasma (anticoagulated), breath, skin, urine, feces, and others).
- Sampling extraction technique (if applicable). For breath, if it has been stored in Tedlar® bags, canisters, desorption tubes, SPME, etc. For serum, the time and temperature allowed for clotting. For both serum and plasma the temperature, time, and speed of centrifugation. For *post mortem* tissues, hours after death, storage conditions, etc.
- Volume collected.
- The frequency of collection (if applicable).
- Storage temperature (freezing?) and time.
- Transport conditions (e.g., inside dry ice) and time.
- Identification of samples (if coded was used).

IV. Instrumentation

- Instrument descriptor (manufacturer, number, software package and version number and/or date). For GC-MS, provide this information for both GC and MS.
- Instrument characteristics. For GC-MS, provide detailed information for GC (auto-injector, separation column, and pre/guard column, separation parameters) and MS (sample introduction and delivery,

an ionization source, mass analyzer description and acquisition mode, data acquisition parameters, internal calibrant(s) if relevant).
- The volume injected/analyzed.
- Technique specific sample preparation (if applicable). For example, sample spiking with internal standards or retention index standards.
- Instrumentation performance (calibration used) and method used for validation (e.g., internal standards). This will ensure reliable and reproducible data production.

V. Data pre-processing
- Data file format used and/or conversion methods.
- Pre-processing methods which convert raw instrumental data into organized or tabular file formats. For example, in MS it will include background subtraction, noise reduction, curve resolution for temporal chromatographic alignment, peak picking, peak thresholding, spectral deconvolution and/or metabolite identifications.

VI. Metabolite identification (if applicable)
- Authors should clearly differentiate and report the level of identification rigor for all metabolites reported.
- Level 1 identification requires that two or more orthogonal properties are reported and that they were obtained comparing the experimental data with authentic chemical standard analyzed in the same laboratory with the same analytical methods.
- Levels 2 and 3 do not require matching to data from authentic chemical standards acquired within the same laboratory. However, report for the annotated ones (level 2) a minimum of one chemical name (IUPAC or common) and one structural code (SMILES or InChI).
- Levels 3 and 4 compounds provide the retention time and/or main ions in the mass spectrum.

1.6 CONCLUSIONS

Volatilomics is leaving a gold era, where every day more research is done in studying the volatile metabolite fingerprint from biological matrices because easy collectible samples (mainly breath and urine) are being used for disease diagnosis, stratification, and prognosis. In this sense, volatile metabolomics would be the key in the next years in precision medicine, for the non-invasiveness and the possibility of using sensors as point-of-care devices. The list of human diseases or disorders being interrogated increases everyday, but it has to go hand in hand with the development of informatic tools and with the appearance of the first volatile human volatilome database.

ACKNOWLEDGMENTS

Dr. Cumeras acknowledges the project 2017PMF-POST-10 from the Martí-Franquès Research Programme from the Universitat Rovira I Virgili (URV, Spain).

KEYWORDS

- **data analysis**
- **etiology**
- **metabolomics**
- **precision medicine**
- **standardization**
- **systems biology**
- **volatilome**

REFERENCES

Abaffy, T., Möller, M. G., Riemer, D. D., Milikowski, C., & DeFazio, R. A., (2013). Comparative analysis of volatile metabolomics signals from melanoma and benign skin: a pilot study. *Metabolomics*, *9*(5), 998–1008.

Adams, R. P., (2007). *Identification of Essential Oil Components by Gas Chromatography/ Mass Spectrometry*. Allured Publishing Corporation, Carol Stream.

Agarwal, S. M., Sharma, M., & Fatima, S., (2016). VOCC: A database of volatile organic compounds in cancer. *RSC Advances*, *6*(115), 114783–114789.

Aksenov, A. A., Sandrock, C. E., Zhao, W., Sankaran, S., Schivo, M., Harper, R., Cardona, C. J., Xing, Z., & Davis, C. E., (2014a). Cellular Scent of influenza virus infection. *Chembiochem.*, *15*(7), 1040–1048.

Aksenov, A. A., Yeates, L., Pasamontes, A., Siebe, C., Zrodnikov, Y., Simmons, J., McCartney, M. M., Deplanque, J. P., Wells, R. S., & Davis, C. E., (2014b). Metabolite content profiling of bottlenose dolphin exhaled breath. *Anal. Chem.*, *86*(21), 10616–10624.

Alphus Dan, W., (2016). Biomarker metabolite signatures pave the way for electronic-nose applications in early clinical disease diagnoses. *Curr. Metabolomics*, *4*, 1–12.

Amann, A., De Lacy Costello, B., Miekisch, W., Schubert, J., Buszewski, B., Pleil, J., Ratcliffe, N., & Risby, T., (2014). The human volatilome: volatile organic compounds (VOCs) in exhaled breath, skin emanations, urine, feces and saliva. *J. Breath Res.*, *8*(3), 034001.

Ashrafi, M., Bates, M., Baguneid, M., Alonso-Rasgado, T., Rautemaa-Richardson, R., & Bayat, A., (2017). Volatile organic compound detection as a potential means of diagnosing cutaneous wound infections. *Wound Repair Regen.*, *25*(4), 574–590.

Baldwin, I. T., Halitschke, R., Paschold, A., Von Dahl, C. C., & Preston, C. A., (2006). Volatile signaling in plant-plant interactions: "Talking trees" in the genomics era. *Science*, *311*(5762), 812–815.

Banday, K. M., Pasikanti, K. K., Chan, E. C. Y., Singla, R., Rao, K. V. S., Chauhan, V. S., & Nanda, R. K., (2011). Use of urine volatile organic compounds to discriminate tuberculosis patients from healthy subjects. *Anal. Chem.*, *83*(14), 5526–5534.

Beale, D., Jones, O., Karpe, A., Dayalan, S., Oh, D., Kouremenos, K., Ahmed, W., & Palombo, E., (2017). A review of analytical techniques and their application in disease diagnosis in breathomics and salivaomics research. *Int. J. Mol. Sci.*, *18*(1), 24.

Bekri, S., (2016). The role of metabolomics in precision medicine. *Expert Rev. Precis. Med. Drug*, *1*(6), 517–532.

Berkhout, D., Benninga, M., Van Stein, R., Brinkman, P., Niemarkt, H., De Boer, N., & De Meij, T., (2016). Effects of sampling conditions and environmental factors on fecal volatile organic compound analysis by an electronic nose device. *Sensors*, *16*(11), 1967.

Berna, A. Z., McCarthy, J. S., Wang, R. X., Saliba, K. J., Bravo, F. G., Cassells, J., Padovan, B., & Trowell, S. C., (2015). Analysis of breath specimens for biomarkers of plasmodium falciparum infection. *J. Infect. Dis.*, *212*(7), 1120–1128.

Blanchet, L., Smolinska, A., Baranska, A., Tigchelaar, E., Swertz, M., Zhernakova, A., Dallinga, J. W., Wijmenga, C., & Van Schooten, F. J., (2017). Factors that influence the volatile organic compound content in human breath. *J. Breath Res.*, *11*(1), 016013.

Bos, L. D. J., Sterk, P. J., & Schultz, M. J., (2013). Volatile metabolites of pathogens: a systematic review. *PLoS Pathog.*, *9*(5), e1003311.

Bouza, M., Gonzalez-Soto, J., Pereiro, R., De Vicente, J. C., & Sanz-Medel, A., (2017). Exhaled breath and oral cavity VOCs as potential biomarkers in oral cancer patients. *J. Breath Res.*, *11*(1), 016015.

Cauchi, M., Fowler, D. P., Walton, C., Turner, C., Jia, W., Whitehead, R. N., Griffiths, L., Dawson, C., Bai, H., Waring, R. H., Ramsden, D. B., Hunter, J. O., Cole, J. A., & Bessant, C., (2014). Application of gas chromatography mass spectrometry (GC–MS) in conjunction with multivariate classification for the diagnosis of gastrointestinal diseases. *Metabolomics*, *10*(6), 1113–1120.

Cavaleiro Rufo, J., Madureira, J., Oliveira Fernandes, E., & Moreira, A., (2016). Volatile organic compounds in asthma diagnosis: a systematic review and meta-analysis. *Allergy*, *71*(2), 175–188.

Chouinard, C. D., Wei, M. S., Beekman, C. R., Kemperman, R. H. J., & Yost, R. A., (2016). Ion mobility in clinical analysis: Current progress and future perspectives. *Clin. Chem.*, *62*(1), 124–133.

Christiaan, A. R., Agnieszka, S., & Jane, E. H., (2016). The volatile metabolome of *Klebsiella pneumoniae* in human blood. *J. Breath Res.*, *10*(2), 027101.

Covington, J. A., Van. der Schee, M. P., Edge, A. S. L., Boyle, B., Savage, R. S., & Arasaradnam, R. P., (2015). The application of FAIMS gas analysis in medical diagnostics. *Analyst.*, *140*(20), 6775–6781.

Cumeras, R., (2017). Volatilome metabolomics and databases, recent advances and needs. *Curr. Metabolomics*, *5*, 79–89.

Cumeras, R., Aksenov, A. A., Pasamontes, A., Fung, A. G., Cianchetta, A. N., Doan, H., Davis, R. M., & Davis, C. E., (2016). Identification of fungal metabolites from inside *Gallus gallus* domesticus eggshells by non-invasively detecting volatile organic compounds (VOCs). *Anal. Bioanal. Chem.*, *408*(24), 6649–6658.

Cumeras, R., Cheung, W., Gulland, F., Goley, D., & Davis, C., (2014). Chemical analysis of whale breath volatiles: a case study for non-invasive field health diagnostics of marine mammals. *Metabolites*, *4*(3), 790–806.

Cumeras, R., Favrod, P., Buchinger, H., Kreuer, S., Volk, T., Figueras, E., Gràcia, I., Maddula, S., & Baumbach, J. I., (2013). In online-monitoring of drugs with ion mobility spectrometry in patients under anesthesia, *Breath Summit 2013: International Conference of Breath Research*, Saarbrücken/Wallerfangen, Germany, 118.

Daniel, D. A. P., & Thangavel, K., (2016). Breathomics for gastric cancer classification using back-propagation neural network. *J. Med. Signals Sens.*, *6*(3), 172–182.

Das, M. K., Bishwal, S. C., Das, A., Dabral, D., Varshney, A., Badireddy, V. K., & Nanda, R., (2014). Investigation of gender-specific exhaled breath volatome in humans by GCxGC-TOF-MS. *Anal. Chem.*, *86*(2), 1229–1237.

De Angelis, M., Francavilla, R., Piccolo, M., De Giacomo, A., & Gobbetti, M., (2015). Autism spectrum disorders and intestinal microbiota. *Gut Microbes*, *6*(3), 207–213.

De Boer, N. K. H., De Meij, T. G. J., Oort, F. A., Ben Larbi, I., Mulder, C. J. J., Van Bodegraven, A. A., & Van der Schee, M. P., (2014). The scent of colorectal cancer: Detection by volatile organic compound analysis. *Clin. Gastroenterol. Hepatol.*, *12*(7), 1085–1089.

De Lacy Costello, B., Amann, A., Al-Kateb, H., Flynn, C., Filipiak, W., Khalid, T., Osborne, D., & Ratcliffe, N. M., (2014). A review of the volatiles from the healthy human body. *J. Breath Res.*, *8*(1), 014001.

Di Lena, M., Porcelli, F., & Altomare, D. F., (2016). Volatile organic compounds as new biomarkers for colorectal cancer: A review. *Colorectal Dis.*, *18*(7), 654–663.

Dragonieri, S., Pennazza, G., Carratu, P., & Resta, O., (2017). Electronic nose technology in respiratory diseases. *Lung.*, *195*(2), 157–165.

Drew, L., (2016). Pharmacogenetics: The right drug for you. *Nature*, *537*(7619), S60–S62.

Dudzik, D., Barbas-Bernardos, C., García, A., & Barbas, C., (2018). Quality assurance procedures for mass spectrometry untargeted metabolomics. a review. *J. Pharm. Biomed. Anal.*, *147*, 149–173.

EU CE, (2010), The European Parliament and the Council of the European Union, 2010/75/ EU Directive on industrial emissions (integrated pollution prevention and control). *Official Journal of the European Union*, *334*, 17–119.

Fiehn, O., (2002). Metabolomics – the link between genotypes and phenotypes. *Plant Mol. Biol.*, *48*(1), 155–171.

Filipiak, W., Mochalski, P., Filipiak, A., Ager, C., Cumeras, R., E., Davis, C., Agapiou, A., Unterkofler, K., & Troppmair, J., (2016). A compendium of volatile organic compounds (VOCs) released by human cell lines. *Curr. Med. Chem.*, *23*(20), 2112–2131.

Filipiak, W., Ruzsanyi, V., Mochalski, P., Filipiak, A., Bajtarevic, A., Ager, C., Denz, H., Hilbe, W., Jamnig, H., Hackl, M., Dzien, A., & Amann, A., (2012). Dependence of exhaled breath composition on exogenous factors, smoking habits and exposure to air pollutants. *J. Breath Res.*, *6*(3), 036008.

Fitzgerald, J. E., Bui, E. T. H., Simon, N. M., & Fenniri, H., (2017). Artificial nose technology: Status and prospects in diagnostics. *Trends Biotechnol.*, *35*(1), 33–42.

Francavilla, R., Ercolini, D., Piccolo, M., Vannini, L., Siragusa, S., De Filippis, F., De Pasquale, I., Di Cagno, R., Di Toma, M., Gozzi, G., Serrazanetti, D. I., De Angelis, M., & Gobbetti, M., (2014). Salivary microbiota and metabolome associated with celiac disease. *Appl. Environ. Microbiol.*, *80*(11), 3416–3425.

Gałęzowska, G., Chraniuk, M., & Wolska, L., (2016). *In vitro* assays as a tool for determination of VOCs toxic effect on respiratory system: A critical review. *Trends Analyt. Chem.*, *77*, 14–22.

Gardner, J., & Vincent, T., (2016). Electronic noses for well-being: Breath analysis and energy expenditure. *Sensors*, *16*(7), 947.

Gibney, M. J., Walsh, M., Brennan, L., Roche, H. M., German, B., & Van Ommen, B., (2005). Metabolomics in human nutrition: Opportunities and challenges. *Am. J. Clin. Nutr.*, *82*(3), 497–503.

Gruber, M., Tisch, U., Jeries, R., Amal, H., Hakim, M., Ronen, O., Marshak, T., Zimmerman, D., Israel, O., Amiga, E., Doweck, I., & Haick, H., (2014). Analysis of exhaled breath for diagnosing head and neck squamous cell carcinoma: A feasibility study. *Br. J. Cancer.*, *111*(4), 790–798.

Guerrero-Flores, H., Apresa-García, T., Garay-Villar, Ó., Sánchez-Pérez, A., Flores-Villegas, D., Bandera-Calderón, A., García-Palacios, R., Rojas-Sánchez, T., Romero-Morelos, P., Sánchez-Albor, V., Mata, O., Arana-Conejo, V., Badillo-Romero, J., Taniguchi, K., Marrero-Rodríguez, D., Mendoza-Rodríguez, M., Rodríguez-Esquivel, M., Huerta-Padilla, V., Martínez-Castillo, A., Hernández-Gallardo, I., López-Romero, R., Bandala, C., Rosales-Guevara, J., & Salcedo, M., (2017). A non-invasive tool for detecting cervical cancer odor by trained scent dogs. *BMC Cancer*, *17*(1), 79.

Guo, L., Wang, C., Chi, C., Wang, X., Liu, S., Zhao, W., Ke, C., Xu, G., & Li, E., (2015). Exhaled breath volatile biomarker analysis for thyroid cancer. *Transl. Res.*, *166*(2), 188–195.

Haggarty, J., & Burgess, K. E. V., (2017). Recent advances in liquid and gas chromatography methodology for extending coverage of the metabolome. *Curr. Opin. Biotechnol.*, *43*, 77–85.

Haick, H., Broza, Y. Y., Mochalski, P., Ruzsanyi, V., & Amann, A., (2014). Assessment, origin, and implementation of breath volatile cancer markers. *Chem. Soc. Rev.*, *43*(5), 1423–1449.

Haitham, A., Marcis, L., Yoav, Y. B., Ulrike, T., Konrads, F., Inta, L. K., Roberts, S., Zhen-qin, X., Hu, L., & Hossam, H., (2013). Geographical variation in the exhaled volatile organic compounds. *J. Breath Res.*, *7*(4), 047102.

Haug, K., Salek, R. M., Conesa, P., Hastings, J., De Matos, P., Rijnbeek, M., Mahendraker, T., Williams, M., Neumann, S., Rocca-Serra, P., Maguire, E., González-Beltrán, A., Sansone, S. A., Griffin, J. L., & Steinbeck, C., (2013). MetaboLights–an open-access general-purpose repository for metabolomics studies and associated meta-data. *Nucleic Acids Res.*, *41*(D1), D781–D786.

Heaney, L. M., & Lindley, M. R., (2017). Translation of exhaled breath volatile analyses to sport and exercise applications. *Metabolomics*, *13*(11), 139.

Heddergott, C., Calvo, A. M., & Latgé, J. P., (2014). The volatome of *Aspergillus fumigatus*. *Eukaryot. Cell.*, *13*(8), 1014–1025.

Heil, M., (2014). Herbivore-induced plant volatiles: targets, perception and unanswered questions. *New Phytol.*, *204*(2), 297–306.

Hochmuth, D., (2017). Mass Finder 4. http://massfinder.com/wiki/MassFinder_4 (accessed February 28, 2017).

Horai, H., Arita, M., Kanaya, S., Nihei, Y., Ikeda, T., Suwa, K., Ojima, Y., Tanaka, K., Tanaka, S., Aoshima, K., Oda, Y., Kakazu, Y., Kusano, M., Tohge, T., Matsuda, F., Sawada, Y., Hirai, M. Y., Nakanishi, H., Ikeda, K., Akimoto, N., Maoka, T., Takahashi, H., Ara, T., Sakurai, N., Suzuki, H., Shibata, D., Neumann, S., Iida, T., Tanaka, K., Funatsu, K., Matsuura, F., Soga, T., Taguchi, R., Saito, K., & Nishioka, T., (2010). Mass Bank: a public repository for sharing mass spectral data for life sciences. *J. Mass Spectrom.*, *45*(7), 703–714.

Horváth, I., Lázár, Z., Gyulai, N., Kollai, M., & Losonczy, G., (2009). Exhaled biomarkers in lung cancer. *Eur. Respir. J.*, *34*(1), 261–275.

Huang, Z., Zhang, J., Zhang, P., Wang, H., Pan, Z., & Wang, L., (2016). Analysis of volatile organic compounds in pleural effusions by headspace solid-phase microextraction coupled with cryotrap gas chromatography and mass spectrometry. *J. Sep. Sci.*, *39*(13), 2544–2552.

Ideker, T., Galitski, T., & Hood, L., (2001). A new approach to decoding life: Systems biology. *Annu. Rev. Genomics Hum. Genet.*, *2*, 343–372.

Jobu, K., Sun, C., Yoshioka, S., Yokota, J., Onogawa, M., Kawada, C., Inoue, K., Shuin, T., Sendo, T., & Miyamura, M., (2012). Metabolomics study on the biochemical profiles of odor elements in urine of human with bladder cancer. *Biol. Pharm. Bull.*, *35*(4), 639–642.

Junker, R. R., & Tholl, D., (2013). Volatile organic compound mediated interactions at the plant-microbe interface. *J. Chem. Ecol.*, *39*(7), 810–825.

Kędziora, K., & Wasiak, W., (2017). Extraction media used in needle trap devices—Progress in development and application. *J. Chromatogr. A.*, *1505*, 1–17.

Kesselmeier, J., & Staudt, M., (1999). Biogenic volatile organic compounds (VOC): An overview on emission, physiology and ecology. *J. Atmos. Chem.*, *33*(1), 23–88.

Kessler, A., & Baldwin, I. T., (2001). Defensive function of herbivore-induced plant volatile emissions in nature. *Science*, *291*(5511), 2141–2144.

Kim, K. H., Szulejko, J. E., Kwon, E., & Deep, A., (2016). A critical review on the diverse preconcentration procedures on bag samples in the quantitation of volatile organic compounds from cigarette smoke and other combustion samples. *Trends Analyt. Chem.*, *85*(Part C), 65–74.

Kind, T., Wohlgemuth, G., Lee, D. Y., Lu, Y., Palazoglu, M., Shahbaz, S., & Fiehn, O., (2009). FiehnLib – mass spectral and retention index libraries for metabolomics based on quadrupole and time-of-flight gas chromatography/mass spectrometry. *Anal. Chem.*, *81*(24), 10038–10048.

Ko, P. W., Kang, K., Yu, J. B., Huh, J. S., Lee, H. W., & Lim, J. O., (2014). Breath gas analysis for a potential diagnostic method of neurodegenerative diseases. *Sens. Lett.*, *12*(6–7), 1198–1202.

Koo, S., Thomas, H. R., Daniels, S. D., Lynch, R. C., Fortier, S. M., Shea, M. M., Rearden, P., Comolli, J. C., Baden, L. R., & Marty, F. M., (2014). A breath fungal secondary metabolite signature to diagnose invasive *Aspergillosis*. *Clin. Infect. Dis.*, *59*(12), 1733–1740.

Kopka, J., Schauer, N., Krueger, S., Birkemeyer, C., Usadel, B., Bergmüller, E., Dörmann, P., Weckwerth, W., Gibon, Y., Stitt, M., Willmitzer, L., Fernie, A. R., & Steinhauser, D., (2005). GMD@CSB. DB: the Golm metabolome database. *Bioinformatics*, *21*(8), 1635–1638.

Kurada, S., Alkhouri, N., Fiocchi, C., Dweik, R., & Rieder, F., (2015). Review article: breath analysis in inflammatory bowel diseases. *Aliment. Pharmacol. Ther.*, *41*(4), 329–341.

Lawal, O., Ahmed, W. M., Nijsen, T. M. E., Goodacre, R., & Fowler, S. J., (2017). Exhaled breath analysis: a review of 'breath-taking' methods for off-line analysis. *Metabolomics*, *13*(10), 110.

Lemfack, M. C., Nickel, J., Dunkel, M., Preissner, R., & Piechulla, B., (2014). mVOC: a database of microbial volatiles. *Nucleic Acids Res.*, *42*(Database issue), 744–748.

Li, N., Alfiky, A., Vaughan, M. M., & Kang, S., (2016). Stop and smell the fungi: Fungal volatile metabolites are overlooked signals involved in fungal interaction with plants. *Fungal Biol. Rev.*, *30*(3), 134–144.

Li, W., Liu, H. Y., Jia, Z. R., Qiao, P. P., Pi, X. T., Chen, J., & Deng, L. H., (2014). Advances in the early detection of lung cancer using analysis of volatile organic compounds: From imaging to sensors. *Asian Pac. J. Cancer Prev.*, *15*(11), 4377–4384.

Lima, A. R., Bastos, M. D. L., Carvalho, M., & Guedes de Pinho, P., (2016). Biomarker discovery in human prostate cancer: An update in metabolomics studies. *Transl. Oncol.*, *9*(4), 357–370.

Lubes, G., & Goodarzi, M., (2018). GC–MS based metabolomics used for the identification of cancer volatile organic compounds as biomarkers. *J. Pharm. Biomed. Anal.*, *147*, 313–322.

Luque de Castro, M. D., & Priego-Capote, F., (2018). The analytical process to search for metabolomics biomarkers. *J. Pharm. Biomed. Anal.*, *147*(Supplement C), 341–349.

Ma, S., Yim Sun, H., Lee, S. G., Kim Eun, B., Lee, S. R., Chang, K. T., Buffenstein, R., Lewis, K. N., Park, T. J., Miller, R. A., Clish, C. B., & Gladyshev, V. N., (2015). Organization of the mammalian metabolome according to organ function, lineage specialization, and longevity. *Cell. Metab.*, *22*(2), 332–343.

Manig, F., Kuhne, K., Von Neubeck, C., Schwarzenbolz, U., Yu, Z., Kessler, B. M., Pietzsch, J., & Kunz-Schughart, L. A., (2017). The why and how of amino acid analytics in cancer diagnostics and therapy. *J. Biotechnol.*, *242*, 30–54.

Markar, S. R., Wiggins, T., Kumar, S., & Hanna, G. B., (2015). Exhaled breath analysis for the diagnosis and assessment of endoluminal gastrointestinal diseases. *J. Clin. Gastroenterol.*, *49*(1), 1–8.

Menon, S. S., Uppal, M., Randhawa, S., Cheema, M. S., Aghdam, N., Usala, R. L., Ghosh, S. P., Cheema, A. K., & Dritschilo, A., (2016). Radiation metabolomics: Current status and future directions. *Front. Oncol.*, *6*, 20.

Michael, P., Renee, N. C., Christobel, S., Peter, H., Peter, S., & James, W., (2010). Volatile biomarkers in the breath of women with breast cancer. *J. Breath Res.*, *4*(2), 026003.

Michael, R. P., Bonsall, R. W., & Warner, P., (1974). Human vaginal secretions: Volatile fatty acid content. *Science*, *186*(4170), 1217–1219.

Miller, G. W., & Jones, D. P., (2014). The nature of nurture: Refining the definition of the exposome. *Toxicol. Sci.*, *137*(1), 1–2.

Milman, B. L., (2015). General principles of identification by mass spectrometry. *Trends Analyt. Chem.*, *69*, 24–33.

Milman, B. L., & Zhurkovich, I. K., (2016). Mass spectral libraries: A statistical review of the visible use. *Trends Analyt. Chem.*, *80*, 636–640.

Minai-Tehrani, A., Jafarzadeh, N., & Gilany, K., (2016). Metabolomics: a state-of-the-art technology for better understanding of male infertility. *Andrologia.*, *48*(6), 609–616.

Mirzaei, A., Leonardi, S. G., & Neri, G., (2016). Detection of hazardous volatile organic compounds (VOCs) by metal oxide nanostructures-based gas sensors: A review. *Ceram. Int.*, *42*(14), 15119–15141.

Muñoz-Hernández, A. M., Galbis-Estrada, C., Santos-Bueso, E., Cuiña-Sardiña, R., Díaz-Valle, D., Gegúndez-Fernández, J. A., Pinazo-Durán, M. D., & Benítez-del-Castillo, J. M., (2016). Human tear metabolome. *Archivos de la Sociedad Española de Oftalmología* (English Edition), *91*(4), 157–159.

Nakhleh, M. K., Quatredeniers, M., & Haick, H., (2017). Detection of halitosis in breath: Between the past, present, and future. *Oral Dis.*, *00*, 1–11.

Navaneethan, U., Parsi, M. A., Gutierrez, N. G., Bhatt, A., Venkatesh, P. G. K., Lourdusamy, D., Grove, D., Hammel, J. P., Jang, S., Sanaka, M. R., Stevens, T., Vargo, J. J., & Dweik, R. A., (2014). Volatile organic compounds in bile can diagnose malignant biliary strictures in the setting of pancreatic cancer: A preliminary observation. *Gastrointest. Endosc.*, *80*(6), 1038–1045.

Neerincx, A. H., Vijverberg, S. J. H., Bos, L. D. J., Brinkman, P., Van der Schee, M. P., De Vries, R., Sterk, P. J., & Maitland-van der Zee, A.-H., (2017). Breathomics from exhaled volatile organic compounds in pediatric asthma. *Pediatr. Pulmonol.*, *52*(12), 1616–1627.

NIH, (2017). National Institutes of Health All of Us Research Program. https://www. nih. gov/research-training/allofus-research-program (accessed 20 Feb, 2017).

NIST, (2017). 14 Mass Spectral Library. GC Methods/Retention Index Database. http:// nistmassspeclibrary. com/nist-spectra-gc-methodsretention-index-database/ (accessed 28 Feb, 2017).

Nizio, K. D., Perrault, K. A., Troobnikoff, A. N., Shoma, S., Iredell, J. R., Middleton, P. G., & Forbes, S. L., (2016). *In vitro* volatile organic compound profiling using GC×GC-TOFMS to differentiate bacteria associated with lung infections: a proof-of-concept study. *J. Breath Res.*, *10*(2), 026008.

Petrof, E. O., Claud, E. C., Gloor, G. B., & Allen-Vercoe, E., (2013). Microbial ecosystems therapeutics: A new paradigm in medicine? *Benef. Microbes.*, *4*(1), 53–65.

Phillips, C., Mac Parthaláin, N., Syed, Y., Deganello, D., Claypole, T., & Lewis, K., (2014). Short-term intra-subject variation in exhaled volatile organic compounds (VOCs) in

COPD patients and healthy controls and its effect on disease classification. *Metabolites*, *4*(2), 300.

Pijnenburg, M. W., & Szefler, S., (2015). Personalized medicine in children with asthma. *Paediatr. Respir. Rev.*, *16*(2), 101–107.

Poulin, P., & Krishnan, K., (1996). A mechanistic algorithm for predicting blood: Air partition coefficients of organic chemicals with the consideration of reversible binding in hemoglobin. *Toxicol. Appl. Pharmacol.*, *136*(1), 131–137.

Prokop-Prigge, K. A., Thaler, E., Wysocki, C. J., & Preti, G., (2014). Identification of volatile organic compounds in human cerumen. *J. Chromatogr. B.*, *953–954*, 48–52.

Robroeks, C. M. H. H. T., Van Berkel, J. J. B. N., Dallinga, J. W., Jobsis, Q., Zimmermann, L. J. I., Hendriks, H. J. E., Wouters, M. F. M., Van der Grinten, C. P. M., Van de Kant, K. D. G., Van Schooten, F.-J., & Dompeling, E., (2010). Metabolomics of volatile organic compounds in cystic fibrosis patients and controls. *Pediatr. Res.*, *68*(1), 75–80.

Saalberg, Y., & Wolff, M., (2016). VOC breath biomarkers in lung cancer. *Clin. Chim. Acta.*, *459*, 5–9.

Schivo, M., Aksenov, A. A., Linderholm, A. L., McCartney, M. M., Simmons, J., Harper, R. W., & Davis, C. E., (2014). Volatile emanations from *in vitro* airway cells infected with human rhinovirus. *J. Breath Res.*, *8*(3), 037110.

Schivo, M., Aksenov, A. A., Pasamontes, A., Cumeras, R., Weisker, S., Oberbauer, A. M., & Davis, C. E., (2017). A rabbit model for assessment of volatile metabolite changes observed from skin: a pressure ulcer case study. *J. Breath Res.*, *11*(1), 016007.

Schmidt, K., & Podmore, I., (2015). Current challenges in volatile organic compounds analysis as potential biomarkers of cancer. *J. Biomark.*, 981458.

Schmidt, R., Cordovez, V., De Boer, W., Raaijmakers, J., & Garbeva, P., (2015). Volatile affairs in microbial interactions. *ISME J.*, *9*(11), 2329–2335.

Schmidt, R., Etalo, D. W., De Jager, V., Gerards, S., Zweers, H., De Boer, W., & Garbeva, P., (2016). Microbial small talk: Volatiles in fungal–bacterial interactions. *Front. Microbiol.*, *6*(1495).

Schnabel, R., Fijten, R., Smolinska, A., Dallinga, J., Boumans, M. L., Stobberingh, E., Boots, A., Roekaerts, P., Bergmans, D., & Van Schooten, F. J., (2015). Analysis of volatile organic compounds in exhaled breath to diagnose ventilator-associated pneumonia. *Sci. Rep.*, *5*, 17179.

Shende, P., Vaidya, J., Kulkarni, Y. A., & Gaud, R. S., (2017). Systematic approaches for biodiagnostics using exhaled air. *J. Control. Release.*, *268*(Supplement C), 282–295.

Silva, C. L., Passos, M., & Camara, J. S., (2011). Investigation of urinary volatile organic metabolites as potential cancer biomarkers by solid-phase microextraction in combination with gas chromatography-mass spectrometry. *Br. J. Cancer.*, *105*(12), 1894–1904.

Skogerson, K., Wohlgemuth, G., Barupal, D. K., & Fiehn, O., (2011). The volatile compound BinBase mass spectral database. *BMC Bioinformatics.*, *12*, 321.

Sornette, D., Yukalov, V. I., Yukalova, E. P., Henry, J. Y., Schwab, D., & Cobb, J. P., (2009). Endogenous versus exogenous origins of diseases. *J. Biol. Syst.*, *17*(02), 225–267.

Sousa, C. M., Biancur, D. E., Wang, X., Halbrook, C. J., Sherman, M. H., Zhang, L., Kremer, D., Hwang, R. F., Witkiewicz, A. K., Ying, H., Asara, J. M., Evans, R. M., Cantley, L. C., Lyssiotis, C. A., & Kimmelman, A. C., (2016). Pancreatic stellate cells support tumor metabolism through autophagic alanine secretion. *Nature*, *536*(7617), 479–483.

Špánik, I., & Machyňáková, A., (2018). Recent applications of gas chromatography with high-resolution mass spectrometry. *J. Sep. Sci.*, *41*(1), 163–179.

Spicer, R. A., Salek, R., & Steinbeck, C., (2017a). Compliance with minimum information guidelines in public metabolomics repositories. *Sci. Data.*, *4*, 170137.

Spicer, R., Salek, R. M., Moreno, P., Cañueto, D., & Steinbeck, C., (2017b). Navigating freely-available software tools for metabolomics analysis. *Metabolomics*, *13*(9), 106.

Sud, M., Fahy, E., Cotter, D., Azam, K., Vadivelu, I., Burant, C., Edison, A., Fiehn, O., Higashi, R., Nair, K. S., Sumner, S., & Subramaniam, S., (2016). Metabolomics workbench: An international repository for metabolomics data and metadata, metabolite standards, protocols, tutorials and training, and analysis tools. *Nucleic Acids Res.*, *44*(D1), 463–470.

Sumner, L. W., Amberg, A., Barrett, D., Beale, M. H., Beger, R., Daykin, C. A., Fan, T. W. M., Fiehn, O., Goodacre, R., Griffin, J. L., Hankemeier, T., Hardy, N., Harnly, J., Higashi, R., Kopka, J., Lane, A. N., Lindon, J. C., Marriott, P., Nicholls, A. W., Reily, M. D., Thaden, J. J., & Viant, M. R., (2007). Proposed minimum reporting standards for chemical analysis. *Metabolomics*, *3*(3), 211–221.

Sun, X., Shao, K., & Wang, T., (2016). Detection of volatile organic compounds (VOCs) from exhaled breath as noninvasive methods for cancer diagnosis. *Anal. Bioanal. Chem.*, *408*(11), 2759–2780.

Tolstikov, V., (2016). Metabolomics: Bridging the gap between pharmaceutical development and population health. *Metabolites*, *6*(3), 20.

Tripathi, K. M., Kim, T., Losic, D., & Tung, T. T., (2016). Recent advances in engineered graphene and composites for detection of volatile organic compounds (VOCs) and non-invasive diseases diagnosis. *Carbon.*, *110*(Supplement C), 97–129.

Trivedi, D. K., Hollywood, K. A., & Goodacre, R., (2017). Metabolomics for the masses: The future of metabolomics in a personalized world. *New Horiz. Transl. Med.*, *3*(6), 294–305.

Tyc, O., Zweers, H., De Boer, W., & Garbeva, P., (2015). Volatiles in inter-specific bacterial interactions. *Front. Microbiol*, *6*(1412).

US EPA, (2017). United States Environmental Protection Agency 40 CFR 51.100(s)-Definition–Volatile organic compounds (VOC). https://archive. epa. gov/ttn/ozone/web/html/def_voc. html (accessed 15 Feb, 2017).

Wang, C., Feng, Y., Wang, M., Pi, X., Tong, H., Wang, Y., Zhu, L., & Li, E., (2015). Volatile organic metabolites identify patients with mesangial proliferative glomerulonephritis, IgA nephropathy and normal controls. *Sci. Rep.*, *5*, 14744.

Wenwen, L., Yong, L., Xiaoyong, L., Yanping, H., Yu, L., Shouquan, C., & Yixiang, D., (2015). A cross-sectional study of breath acetone based on diabetic metabolic disorders. *J. Breath Res.*, *9*(1), 016005.

WHO, (1989), World Health Organization. Indoor air quality: Organic pollutants. Report on a WHO Meeting. *EURO Rep. Stud.*, *111*, 1–70.

Wild, C. P., (2005). Complementing the genome with an "exposome": The outstanding challenge of environmental exposure measurement in molecular epidemiology. *Cancer Epidemiol. Biomarkers Prev.*, *14*(8), 1847–1850.

Wiley Registry® of Mass Spectral Data. http://olabout. wiley. com/WileyCDA/Section/id-406117. html#chemistry (accessed 28 Feb, 2017).

Wilson, A. D., (2017). Biomarker metabolite signatures pave the way for electronic-nose applications in early clinical disease diagnoses. *Curr. Metabolomics*, 5(2), 90–101.

Xue, R., Dong, L., Zhang, S., Deng, C., Liu, T., Wang, J., & Shen, X., (2008). Investigation of volatile biomarkers in liver cancer blood using solid-phase microextraction and gas chromatography/mass spectrometry. *Rapid Commun. Mass Spectrom.*, 22(8), 1181–1186.

Zhang, Q. H., Zhou, L. D., Chen, H., Wang, C. Z., Xia, Z. N., & Yuan, C. S., (2016). Solid-phase microextraction technology for *in vitro* and *in vivo* metabolite analysis. *Trends Analyt. Chem.*, 80, 57–65.

Zhou, J., Huang, Z. A., Kumar, U., & Chen, D. D. Y., (2017). Review of recent developments in determining volatile organic compounds in exhaled breath as biomarkers for lung cancer diagnosis. *Anal. Chim. Acta.*, 996(Supplement C), 1–9.

CHAPTER 2

BASICS OF GAS CHROMATOGRAPHY MASS SPECTROMETRY SYSTEM

WILLIAM HON KIT CHEUNG[1] and RAQUEL CUMERAS[2-4]

[1]*Department of Applied science, Northumbria University, Newcastle upon Tyne, NE1 8ST, UK, Tel.: +44.7468580538, E-mail: William.Cheung@northumbria.ac.uk*

[2]*Metabolomics Platform, Metabolomics Interdisciplinary Laboratory (MiL@b), Department of Electronic Engineering (DEEEA), Universitat Rovira i Virgili, IISPV, 43003 Tarragona, Catalonia, Spain, Tel.: +34.977558764, E-mail: raquel.cumeras@urv.cat*

[3]*Biomedical Research Centre in Diabetes and Associated Metabolic Disorders (CIBERDEM), 28029 Madrid, Spain*

[4]*Signal and Information Processing for Sensing Systems, Institute for Bioengineering of Catalonia (IBEC), Baldiri Reixac 4–8, 08028-Barcelona, Spain*

ABSTRACT

Mass analyzers are varied significantly in their mode of operation and design architecture; the following section provides a brief technical overview of the different types of mass analyzers in use today in relation to gas chromatography (GC) and volatile organic compounds (VOCs) analysis. The first part of the chapters also covers the mechanism of separation involved conventional GC, from idealized plate theory to the Van Deemter Equation, exposing the reader to core concepts in GC. While the second part of the chapter gives an overview of several different mass analyzers and their respective mode of operations, from single quadrupole mass filters (QMF) to triple quadrupoles (QQQ), time of flight (TOF), quadrupole time of flight (qTOF) and Orbitrap Mass analyzers, the description is not mean exhaustive and readers are encouraged to explore further with the reference provided within the chapter.

2.1   INTRODUCTION

When a drop of a liquid containing a mixture of various substances is placed on paper, the liquid begins to spread out on the paper. The various substances within the mixture will spread out at a different rate, however, which gives rise to marks on the paper with different colors. In the 1940s, Richard Synge and Archer Martin (Nobel Prize of Chemistry in 1952) (Synge, 2017) used this and similar phenomena in gas mixtures – for example, to develop different types of chromatography – methods for separating substances in mixtures and for determining the composition of mixtures.

Gas chromatography (GC) definition is include in the Partition chromatography definition, as (IUPAC 2017): *Chromatography in which separation is based mainly on differences between the solubility of the sample components in the stationary phase (GC), or on differences between the solubility's of the components in the mobile and stationary phases (liquid chromatography).*

This chapter introduces the basic core concepts and principles of GC and GC-mass spectrometry (GC-MS) systems. The aim of this chapter is to cover the physical and theoretical aspect of GC separation and the different operating principles of various mass analyzers currently in use for small-scale molecule profiling (metabolomics and volatile organic compounds, VOCs). Readers are encouraged to refer to the references provided for further detailed technical explanations of the principles described in this chapter.

2.2   GAS CHROMATOGRAPHY (GC)

GC is a gas phase method of chromatographic separation based on chemical volatility and is particularly suited to the analysis of VOCs and low molecular weight molecules (Adlard, 2001; Baltussen, 2002; Snow and Slack, 2002; Cevallos-Cevallos, 2009). A schematic of a conventional

GC is depicted in Figure 2.1. The analytical technique exploits the differences in boiling points of the different chemical constituents within a
sample mixture to achieved separation (Grob and Barry, 2004). In GC
analysis the separation takes place within an analytical GC column, a
narrow bore open tubular silica column with functionalized stationary
phase coated onto its inner surface. A small volume of liquid sample
(1–5 µL) is typically injected into the GC system via the injection port
assembly, where the sample is flash vaporized under an inert atmosphere
and deposited onto the head of the analytical column. The sample mixture
will then migrate through the analytical column under the influence of the
carrier gas (mobile phase) and the effect of increasing temperature. As
the sample mixture migrates through the column, individual components
within the sample mixture will be retained at different rates based on its
physiochemical interaction with the analytical column (stationary phase)
and be chromatographically separated as a function of chemical volatility
and temperature of the column. Depending on the requirement of the
chemical analysis and the nature of the sample matrix, different range
of analytical GC columns are available with optimized stationary phase
composition, film thickness and column length compatible for the analysis
of volatiles, fatty acids, pesticides, steroids, amines, and alcohols (Agilent,
2016; Restek, 2016).

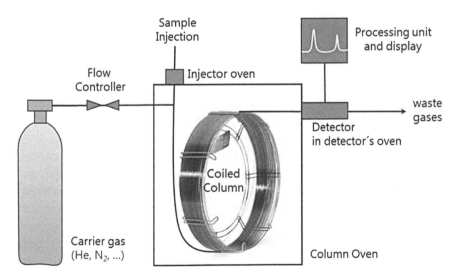

FIGURE 2.1   Gas chromatography schematic.

The migration of the chemical components through the GC column system can be approximated using the plate theory model (James and Martin, 1952), where the GC column can be thought of as a series of theoretical plates sequentially stacked on top of one and another. The chemical component migrates through the GC column by the transfer of equilibrium from one plate under the influence of the carrier gas and temperature.

The resolution equation (Detter-Wlide and Engewald, 2014), which is a quantitative measure of the separation capability of the GC column and is defined as follows:

$$R = (\tfrac{1}{4}\sqrt{N}) \cdot (k/k+1) \cdot (\alpha - 1) \tag{2.1}$$

where R is the baseline resolution ($x \geq 1.5$), $N = L/H$ is the effective number of theoretical plates present, L is the length of the column, $H = HETP$ is the Height Equivalent to Theatrical Plates, k is the retention factor, and α is the selectivity factor.

The resolution equation can be separated into three major components; the first component ($\tfrac{1}{4}\sqrt{N}$) is a measure of column efficiency and is a function of column length, internal column diameter, and the linear velocity of the carrier gas. The second component $(k/k+1)$ is a measure of retention; it is influenced by the internal column diameter, the film thickness of the stationary phase and temperature. The third and final component $(\alpha - 1)$ is a measure of separation, and it is strongly influenced by the type of the stationary phase applied and temperature (Grob and Barry, 2004; Detter-Wlide and Engewald, 2014).

The polarity of the analytical GC column is a function of the stationary phase's chemical composition, and the selection of the appropriate column for a particular application is dictated by both the polarity of the target compounds and the GC column's stationary phase composition. If the stationary phase and the target compounds have comparable polarities, this will result in a strong retention interaction; higher column retention generally equates to greater chromatographic separation and higher resolution. The polarity of the stationary phase will also strongly influence column selectivity and separation. The selectivity factor (α) can be directly related to the stationary phase composition, giving rise to subsequent interaction with the target compounds via intermolecular forces such as hydrogen bonding and dipole–dipole interaction. The effect of temperature is also an important consideration as high polarity stationary phase generally has a lower maximum operating temperature limit (Agilent, 2016; Restek, 2016).

The column film thickness and the internal diameter of the GC column will also strongly influence the retention factor (k) and the resulting resolution. The retention factor is the time required for the target analyte spent in the stationary phase relative to the mobile phase. A combination of low internal column diameter and high film thickness will results in increase retention; however, this effect is attenuated by increasing temperature (Grob and Barry, 2004; Detter-Wlide and Engewald, 2014). For highly volatile compounds increase retention is often required for optimal chromatographic separation, in such instances, it is necessary to use a column with a higher film thickness so that the compound will spend more time in the stationary phase. For high molecular weight compounds, a lower film thickness is preferred as this reduces the length of the time that the compounds are retained by the stationary phase and also reduce the effect of column bleeding at a higher temperature. GC column with a higher film thickness will inherently suffer from a higher rate of column bleed and reduces operating temperature limits.

Low internal diameter GC column will enable higher average linear velocity and also higher number of theoretical plates to be obtained, resulting in sharper peak shape and higher overall separation efficiencies. Low internal diameter column suffers from reduced sample loading capacity due to the lower effective surface area, and adjustment of the sample concentration will be required to account for the lower surface area.

2.2.1   VAN DEEMTER EQUATION

The van Deemter equation (van Deemter, 1956; Detter-Wlide and Engewald, 2014) further built upon the plate model theory, and it is used to accurately describe the various factors which will influence the rate of migration of analytes through the analytical column, and it is expressed as follows:

$$HETP = A + B/U + (C_S + C_M)U \qquad (2.2)$$

where $HETP$ is a measure of resolving power of an analytical column. A is the multiple path or eddy diffusion constant which describe the alternative paths available to the solute within the column stationary phase. A is linearly proportional to the particle size of the stationary phase. The effect of multiple path peak broadening is off-set by the ordinary diffusion at low flow rate typically use in GC system, since the analytes are able to quickly switch between the different paths available thus reducing the time required to move from one plate to another. The effect of multipath diffusion only has a negligible contribution within an open tubular column and GC system. B is the longitudinal diffusion co-efficient, since the concentration of the analyte is lower at the edge of the band relative to its center, the analyte will, therefore, have a tendency to diffuse outwards towards the edge resulting in peak broadening. The effect is inversely proportional to the flow rate. U is the average velocity of the mobile phase (carrier gas) in cm/s^{-1}. C is the mass transfer co-efficient. CS and CM is the mass transfer co-efficient of the stationary and mobile phase respectively. This is the time require for the analytes to move from the mobile phase to the stationary phase and back again. If the velocity of the mobile phase is too high and the analyte is strongly retained by the stationary phase, the analytes in the mobile phases and stationary will traverse the GC column at a vastly different rate resulting in significant peak broadening.

Other factors that contribute to column efficiency are the size of the packing material within the stationary phase and the internal diameter of the GC column itself (previously mentioned). As the lower particle size of the stationary phase will equate to lower stable plates height (lower *HETP*) to be obtained, allowing for an overall increase in the number of theoretical plates per unit column length. The decrease in the internal diameter of the GC column also allows a significant reduction in peak broadening effects as the analytes has a reduced surface area to defuse over as well as higher average linear flow velocity (U) to be obtained.

Within GC systems all the relevant parameters such as flow rate and subsequent heat zones (such as injector port, initial and final GC temperature, ramp rate and MS transferline interface) can be precisely controlled and adjusted as required. The GC system typically uses ultra high purity Helium (H_2) as a carrier gas, which also serves to remove oxygen and other reactive gases from the system. Constant linear flow velocity within the GC column is maintained at all times and across the entire temperature range of the analysis with the use of variable digital mass flow controllers since gas viscosity decreases with increasing temperature. The constant linear flow rate is essential in generating a reproducible chromatographic profile for chemical separation.

As mentioned previously separation within GC is based on the differences in chemical volatility, adjustment of the temperature at which the separation takes place will have a profound effect on the overall chemical analysis. Lower boiling points analytes will traverse the GC column faster and be eluted with respect to higher boiling point analytes. By increasing the column temperature at which the chromatographic separation takes place will results in a net increase in the rate of elution of all components due to greater available thermal energy. Adjustment of the column temperature will strongly affect the overall chromatographic profile (decrease analysis time and improvement of available analytical space); this parameter can be optimized using linear variable temperature programming.

2.2.2   LINEAR VARIABLE TEMPERATURE PROGRAMMING

The linear variable temperature programming (Bakeas, 1996) is a standard practice in both GC and GC-MS analysis; the GC is programmed to increase at a predefine rate over a specific unit of time (the ramp rate and is given in unit of °C/min). This is to exploit inherent differences in the boiling points of different compounds within a sample mixture. This approach has the benefit of reducing the effect of peak broadening, enhanced chromatographic separation, and reduction in the overall analysis time. A lower the ramp rate will enable finer chromatographic separation of compounds with similar boiling points. If the ramp rate is set too shallow however then peak broadening effect would be observed as the chemical components within the sample mixture will migrate through the column extremely slowly, resulting in disproportional increase analysis time and loss in sensitivity. In addition, the peaks will also be too broad and defuse to be integrated and quantified accurately. Conversely, if the temperature ramp rate is set too high, the chromatographic separation will become very crude, as this forces all the components within the sample mixture to migrate through the column at an accelerated rate, resulting in insufficient time for chromatographic separation and poorly resolved peaks. All the constituents within the sample mixture with similar boiling points will be eluted at approximately the same rate, inducing asymmetrical peak shape via co-elution, this will also lead to increased ion suppression effect to be observed. The resulting MS spectra may become nonsensical to interpretation as multiple peaks, and there corresponding MS fragmentation profiles will be overlaid on top of each other in a narrow analytical space.

2.2.3   LOW THERMAL MASS SYSTEM (LTMs)

Low thermal mass system (LTM) (Sloan, 2001; Luong, 2006) is a recent addition to the gas chromatography, whereby the conventional column and associated support control system has been miniaturized to offer equivalent or superior performance in a small, lightweight package. One of the primary advantages of LTM system is that faster and more precise temperature control can be obtained with compare to conventional GC system, as well as overall lower power requirement, this is due to the use of resistive heating and miniaturized column oven, where lower internal volume to heat up and cool down. This in combination with lower internal diameter with a lower film thickness of the stationary phase allows ultra-fast gas phase chromatographic separate to take place.

The LTM system is designed to be hyphenated to MS system to act as a conventional GC front end separation device.

2.2.4   IONIZATION (EI AND CI)

After the sample has been chromatographically separated within the GC system, it is then subjected to ionization prior to MS analysis. The ionization process induces a charge on eluted compounds so that it could be effectively manipulated within the MS system. It is essential for the eluted peak to be as pure as possible prior to the ionization process since the high concentrational uniformity of the eluted peak and it corresponding peak shape helps to reduce the effect of heterogeneous ion suppression. This also improves the overall quality of the corresponding MS data generated, with higher signal-to-noise ratio and reproducible mass fragmentation patterns across the entire peak profile. There are two main methods of sample ionization commonly applied within GC-MS analysis, electron impact ionization (EI) (Märk and Dunn, 1985) and chemical ionization (CI) (Field and Munson, 1965; Todd, 1995). EI is the most routinely used method of ionization in the GC-MS analysis as it offers well-characterized fragmentation profiles. The eluted peak is exposed to a stream of thermionic electrons at 70 eV; inducing large scale fluctuations within the molecule's local electric field, resulting in the removal of a single electron from the highest occupying molecular orbital, generating of a radical cation.

$$M + e^- => M^{+*} + 2e^- \qquad (2.3)$$

The extent of fragmentation will depend on the molecules ionization' cross section area and chemical structure.

However, in some instances electron impact ionization may be too excessive and cause the molecular ion to be present in low abundance or even absent, resulting in inaccurate chemical matching and chemical formula generation. Chemical ionization is a softer form of ionization ($x < 10$ eV), which uses reagent gas such as isobutene, ammonia, methane or carbon dioxide to create a cloud of charge ion to transfer the charge to the analyte (charge transfer reaction). CI does not induce excesses fragmentation due to the low collisional energy used and can be used determine the molecular weight of the parent compound for structure elucidation. CI can be applied in either positive or negative mode polarity; most small scale molecules are able to form positive ions, however, in order to generate a negative ion, the molecule must be able to generate a stable negative charge such as those containing acidic functional groups or halogenated compounds.

2.3 GAS CHROMATOGRAPHY-MASS SPECTROMETRY (GC-MS)

Gas chromatography-mass spectrometry (Dunn, 2005; Skogerson, 2011) is a hyphenated analytical technique that combines the separation capability of the GC with the selectivity and sensitivity of mass spectrometry (MS) for the analysis and quantification of the small-scale molecule. The technique is highly amenable to the analysis of small-scale molecules and VOCs (Robroeks, 2010; Nordström and Lewensohn, 2010; Boots, 2012) due to the nature of separation occurring within the gas phase. The sample is introduced into the GC system where it is separated based on chemical volatility, reducing the complexity of the sample mixture prior to introduction into the MS system for detection and quantification. By simultaneously using both the GC retention time (RT) and MS (mass to charge m/z) information, it enables a high-throughput and robust method of chemical profiling of low molecular weight compounds (Fiehn, 2002; Jonsson, 2005).

2.3.1 MASS SPECTROMETRIC (MS) ANALYZERS

Mass spectrometry is a destructive method of chemical analysis, which provides detailed chemical and structural information regarding the compounds of interest, based on its mass to charge (m/z) ratio and subsequent mass fragmentation patterns; the MS data can be used for structure elucidation and compound identification. Within MS, there are two forms of mass measurements (Kokkonen, 1999) (1) low resolution or unity mass measurement, and (2) accurate or hi-resolution MS measurements. In unity MS measurements, the m/z measurement is reported in the form of the positive integer only (i.e., 121 m/z), whereas, in hi-resolution MS, the m/z measurement is given as a positive integer to 4–5 decimal places (i.e., 121.0457). This difference is non-trivial and is based on the actual atomic mass of the elements itself. Therefore, the more precise the MS measurement, the more accurate the interpretation/inferences can be made regarding the potential elemental composition. Hi resolution MS measurement enables empirical formula to be calculated with greater confidences. The MS patterns of approximately 250,000–400,000 chemical compounds have been characterized and curated within publically available databases such as Human Metabolome Data Base (Wishart, 2012), METLIN (Smith, 2005), Massbank (Horai, 2010), as well as the National institute of standards and technology (NIST) (Halket, 1999).

2.3.2 DIRECT INJECTION MASS ANALYSIS TECHNIQUES

There are multiple names given for this analytical technique, Direct Analysis in Real Time (DART) (Beckman, 2008; Fuhrer, 2011) or Direct Injection Mass spectrometry (DIMS), but the principle is the same; the sample is introduced into the MS system through the use of electrospray ionization interface without any form of chromatographic separation applied, and the entire mass spectrum of the sample is measured and recorded. This is a high-throughput low-level chemical information fingerprinting approach; since no chromatographic separation is applied the complexity of the mixture is not reduced. In DART/DIMS analysis, the effect of ion suppression is significant since the entire sample is being ionized at the same time. If the sample matrix is complex in nature, low abundance ions would likely to be missed due to heterogeneous ionization effect.

2.3.3 QUADRUPOLE MASS FILTER (QMF)

Quadrupole mass filter (QMF) (Dawson, 1976) is one of the simplest and oldest types of mass analyzer in used; the QMF consist of four precisely spaced electrodes collected in pairs. A combination of alternating RF and fix direct current (DC) voltages are applied to the two sets of electrodes where one pair has an applied voltage of $+V_{DC} + V_{RF}\cos(\omega t)$, whilst a $-V_{DC} -V_{RF}\cos(\omega t)$ voltage is applied to the other. The voltages applied are exactly 180° degree out of phase with respect to each other. This results in the creation of a quadrupolar field between the electrodes, which is able to manipulate and focuses the ions as they traverse along the central axis of the QMF cell. The fix DC voltage has the effect of focusing the ions towards the central axis of the QMF cell; whist the alternating RF voltage has the opposite effect, destabilizing the ions away to the central region and towards the electrodes. As the ions enter the QMF region, the mass of the ion as well as its charge and the frequency of the applied electric field will determine the resulting ion trajectories. Only ions with a stable trajectory are able to successfully traverse the QMF and be detected. Ions with an unstable trajectory will eventually collide with the electrodes and be eliminated. The motion of the ions within the QMF can be accurately described by the Mathieu equation (Dawson, 1976), by ramping the RF and DC voltages in a fixed ratio and in increasing amplitude, region of stability are created within the quadruplor field where ions of increasing m/z will have stable ion trajectories and can be sequentially transmitted across the QMF and be detected.

2.3.4 TRIPLE-QUADRUPOLE (QQQ) MASS ANALYZER

The capability of quadrupole base MS analyzers is further expanded upon with the introduction of Triple Quadrupole (QQQ) system (Johnson, 1990), which enables added MS/MS functionality and applications. QQQ system is made of a tandem mass spectrometer, where three quadrupole cells are arranged sequentially (Q_1-Q_2-Q_3, respectively). Depending on how the three quadrupole cells are set up, specific applications such as selective ion monitoring (SIM), selective reaction monitoring (SRM) (Tsugawa, 2014), targeted MS/MS as well conventional full scan profiling can be performed. The principle of operations within a QQQ system is similar to standard QMF. The first cell (Q_1) functions as a conventional mass filter where it sans and select the ions of particular m/z and transmits it to the second quadrupole cell (Q_2). The second QMF cell serves as a collision cell where collision induced disassociation (CID) (Yost, 1979; Douglas, 1982) is applied to create MS/MS fragmentation using collision gas such as Ar or N_2. The resulting product ions are then transmitted to the third and final cell (Q_3) where it scans and quantifies. The Q_2 cell is often replaced with a higher multipole equivalent such as hexapole or octopole QMF cell; this is to allow for higher transmission efficiency and improves resolution at the low mass range.

2.3.5   TIME OF FLIGHT (TOF) MASS ANALYZERS

TOF base MS analyzers are widely applied in the field of metabolomics for the untargeted profiling experiments (Dunn, 2008; Rudnicka, 2011; Skogerson, 2011; Tsugawa, 2011). In TOF mass analyzers (Guilhaus, 1995; Mamyrin, 2001), ions are pulsed into a region known as a flight tube where they are accelerated under known electric field strength within a vacuum. The ions are focused by a series of ion optics and directed towards the detector. Based on the time required for the ions to traverse the length of flight tube (the transient time) and be detected, the original mass of the ion can be determined accurately. As the packet of ions travels towards the detector, it may become defocused or spread out due to acceleration effect of the electric field over a large macromolecular distance, resulting in reduced mass resolution and sensitivity. In order to off-set the spread of the ions, a reflectron system is incorporated into all modern TOF system. A reflectron system serves two main functions: (1) to focus and redirect the flight path of the ions towards the detector, and (2) increase the distance available within the flight tube without significantly increasing its size. The longer available flight path created by the refectron system also has the added benefit of enhancing the overall mass resolution. A single stage reflectron consists of a series of stack electrodes generating a static homogeneous electric field, as the packet of ions encounters the reflectron, slight differences in the acceleration and spread of the incoming ion is adjusted and refocused. Ions of different acceleration energy will impinge into the reflectron's electric field at varying degree; higher energy ions will penetrate further into the reflectron electric field before being redirected outwards relative to lower energy ions of the same packet. As a result, all the ions within the same packet will exit the reflectron as a more spatially uniform packet and be directed towards the detector. Single and dual-stage reflectron (Mamyrin, 2001) systems have been developed to greatly enhanced the mass range and mass accuracy of TOF base MS system. In order for TOF mass analyzer to generate reproducible and accurate mass measurements, it is critical that the system is properly calibrated prior to analysis. The mass accuracy of the system is constantly being monitored and adjusted by continuously infusing a reference solution of known composition and concentration into the system (lock mass solution).

2.3.6   QUADRUPOLE TIME OF FLIGHT MASS SPECTROMETRY (qTOF) MASS ANALYZERS

The qTOF mass analyzer is another form of tandem MS system (Ens, 2005; Chernushevich, 2001), which combines the functionality of both quadrupole and TOF mass analyzers into a single platform. The operating principle of both quadrupole and TOF mass analyzer has already been previously discussed above. The qTOF mass analyzer uses the same operation principle as TOF system but is provides added flexibility for MS/MS analysis to be performed and provides enhances sensitivity, mass resolution, and mass accuracy. The frontal quadrupole cell operates as ion guide in profiling mode and as a mass selector in MS/MS mode. A multi-pole collision cell placed between the quadrupole cell and the TOF mass analyzer to induced ion fragmentation in MS/MS experiments. Finally, the TOF system acts as the primary mass analyzer in both full scan and MS/MS analysis.

2.3.7   ORBITRAPS MASS ANALYZERS

Mass spectrometry is unique in some sense where the different types of mass analyzers have the inherently different principle of operations, each with its own advantages and disadvantages. For example, quadrupole base systems are essentially RF mass filters, which offers ease of operation, robustness and reasonable scan speeds but are limited to low mass range applications. Fourier transforms ion cyclotrons resonance (FT-ICR) system (not described here) has ultra-high resolving power $x > 500,000$ and mass accuracy, good dynamic range but suffers from low ion transmission efficiency, limited scan speed and are complicated to operate and maintain. Whereas ion trap base systems (also not describe here) offers simplicity in design, high ion transmission and trapping capabilities with reasonable dynamic range and scans speed, but are limited in resolving power and mass accuracy. Time-of-flight systems offers have reasonable mass resolving power, scan rate, and ion detections, but are limited to low ion transmission efficiency and limited dynamic range. Orbitraps mass system takes the best qualities of each type of mass analyzers and integrates them into radically new design architecture. The oribitrap system (Hu, 2005; Makarov, 2006; Perry, 2008; Peterson, 2010) incorporated the use of hyper precision shaped electrodes design from ion trap mass system, image current detection from FT-ICR and TOF based pulse ion injections into electrostatic fields and integrated those principles into a single analytical platform to provided high through-put and ultra high resolution MS analysis.

The *orbitrap* system design is a radical departure in MS analyzer, after GC separation the eluents are introduced into the customized ion source system (extractabrite ion source) a dual ionization source capable of performing either electron impact or chemical ionization. For full scan operation first the ions are the guided by a bent flatapole ion transmission assembly which has a 90° bent curvature prior to the quadrupole interface, this prevents neutral molecules from entering quadrupole mass filter; as neutral molecule are not influence by the electric field of the focusing ion optics and are subsequently collided with the wall of the flatapole beam guild manifold and be eliminated. The ions are then introduced into a quad-rupole mass filter, which operates with an isolation width of 0.4 Daltons, this narrow isolation width enhances overall sensitively and selectively of ions. The ions are then guided pass the C-trap (a curve linear ion trap system), which traps and focus the ions for precise ion injection into the orbitrap mass analyzer.

Finally, a second stage ion optics focus and guides the ions from the C-trap into orbitrap mass analyzer, as the ions enter the mass analyzer, an increasing voltage is applied to the central spindle electrode. This induces the initial packet of ions to be squeezed towards the electrodes and osculated around it in an orbital motion, as the voltage to the central electrode increases, the radius of ion trajectory decreases due to an effect know as electrodynamics squeezing. The applied voltages ramp is then ceased, and the ion osculation is stabilized and trapped around the central spindle electrode. Ions of different *m/z* will osculate at independent orbital trajectories with respect to other. As the ions osculate back and forth around the central electrodes, a sinusoidal signal is produced and detected using a differential amplifier. For multiple signals, the super positioning of multiple sinusoidal signals are observed, the different frequencies of the sinusoidal signals equate to the different m/z and are inferred using inverse FT signal transformation.

By combining the ion trap confinement principle for the trapping and transmission of ions, with quadrupole mass selectivity for ion filtering in combination with the ultra-high resolving power of the orbitap mass analyzer, orbitrap MS represents the current gold standard in MS analysis for compound identifications and annotations within small scale analysis.

2.3.8 POST DATA ACQUISITION AND PROCESSING

Once the data have been acquired and evaluated to be of sufficient quality; the data is processed for alignment and peak table generation. One of the first steps is deconvolution of the GC-MS data set, were by the individual ions are extracted from the total ion chromatogram, even at trace levels abundances, parsed according to retention times and m/z values according to specified tolerances (signal to noise ratio, background noise level, retention time windows and mass tolerances). One of the most well-known programs for the processing of GC-MS data is automated mass spectral deconvolution and identification system (AMDIS) (Stein, 1999). The de-convoluted data is aligned and parsed accordingly to create a peak table where the entire MS feature detected within all the samples along with it corresponding fragmentation pattern and associated retention time. One of the most used alignment engines is spectconnect (www.spectconnect.mit.edu) (Styczynski, 2007). The two software solution mentioned above predominately only work for unity type data set, but both software is compatible with most if not all data format from different instrument vendors, for Hi-resolution GC-MS data each instrument vendor will its own proprietary software solution.

2.4 CONCLUSIONS

GC-MS is a relatively mature method in metabolomics because the reproducible measurement is possible and numerous peaks can be reliably obtained from a biological sample. In addition, peak identification is straightforward when RT and mass spectra data are compared to those of accumulated compound information (reference library). For these reasons, GC-MS is generally recognized as one of the most versatile and applicable platforms in metabolomics.

ACKNOWLEDGMENTS

Dr. Cumeras acknowledges the project 2017PMF-POST-10 from the Martí-Franquès Programme (URV, Spain).

KEYWORDS

- **GC**
- **GC-MS**
- **mass analyzers**
- **metabolomics**
- **Van Deemeter equation**
- **VOCs**

REFERENCES

Agilent, (2016), Agilent J&W GC column selection guide. http://www.agilent.com/cs/library/catalogs/public/5990–9867EN_GC_CSG/pdf (accessed Jun 21, 2016).

Bakeas, E. B., & Siskos, P. A., (1996). Effect of temperature programming and pressure on separation number and height equivalent to a theoretical plate in optimization of a serially coupled, open-tubular columns gas chromatographic system. *Anal. Chem.*, *68*, 4468–4473.

Baltussen, E., Cramers, C., & Sandra, P., (2002). Sorptive sample preparation – A review. *Anal. Bioanal. Chem.*, *373*, 3–22.

Beckman, M., Parker, D., Enot, P. E., Duval, E., & Draper, J., (2008). High-throughput, non-targeted metabolite fingerprinting using nominal mass flow infection electrospray mass spectrometry. *Nat. Protoc.*, *3*, 486–504.

Boots, A. W., Berkel, J. J. B. N., Dallinga, J. W., Smolinska, A., Wouters, E. F., & Van Schooten, F. J., (2012). The versatile use of exhaled volatile organic compounds in human health and disease. *J. Breath Res.*, *6*, 027108.

Cevallos-Cevallos, J. M., Reyes-De-Corcuera, J. I., Etxeberria, E., Danyluk, M. D., & Rodrick, G. E., (2009). Metabolomic analysis in food science: a review. *Trends Food Sci. Technol.*, *20*, 557–566.

Chernushevich, I. V., Lobida, A. V., & Thomson, B. A., (2001). An introduction to quadrupole time of flight mass spectrometry. *J. Mass Spectrom.*, *36*, 849–865.

Dawson, P. H., (1976). *Quadrupole Mass Spectrometry and Its Applications, 1ˢᵗ edn.* Elsevier Scientific Company: Amsterdam-Oxford, New York, eBook ISBN: 9781483165042, pp. 372.

Dettmer, K., Aronov, P., & Hammock, B. D., (2007). Mass spectrometry based metabolomics. *Mass Spectrom. Rev.*, *26*, 51–78.

Dettmer-Wlide, K., & Engewald, W., (2014). *Practical Gas Chromatography: A Comprehensive Reference, 1st edn.* Springer, Berlin, Heidelberg, eBook ISBN: 978-3-642-54640-2, pp. 902.

Douglas, D. J., (1982). Mechanism of collision-induced dissociation of polyatomic studied by triple quadrupole mass spectrometry. *J. Phys. Chem.*, *86*, 185–191.

Dunn, W. B., & Ellis, I. D., (2005). Metabolomics: current analytical platform and methodologies. *Trends Anal. Chem.*, *24*, 285–294.

Dunn, W. B., Broadhurst, D., Ellis, D. I., Brown, M., Halsall, A., O'Hagan, S., Spasic, I., Tseng, A., & Kell, B. D., (2008). A GC-TOF-MS study of the stability of serum and urine metabolome during the UK Biobank sample collection and preparation protocols. *Int. J. Epidemiol.*, *37*, 23–30.

Ens, W., & Standing, K. G., (2005). Hybrid quadrupole/time-of-flight mass spectrometers for analysis of biomolecules. *Methods Enzymol.*, *402*, 49–78.

Falkovich, A. H., & Rudich, Y., (2001). Analysis of semi volatile organic compounds in atmospheric aersols by direct sample introduction thermal desorption GC-MS. *Environ. Sci. Technol.*, *35*, 2326–2333.

Fiehn, O., (2002). Metabolome-the link between genotypes and phenotypes. *Plant Mol. Biol.*, *48*, 155–171.

Field, F. H., & Munson, M. S. B., (1965). Reaction of gaseous ions. XIV. Mass spectrometric studies of methane at pressure of 2 Torr. *J. Am. Chem. Soc.*, *87*, 3289–3294.

Fuhrer, T., Heer, D., Begemann, B., & Zamboni, N., (2011). High-thoughput accurate mass metabolome profiling of cellular extract by flow injection time of flight mass spectrometry. *Anal. Chem.*, *83*, 7074–7080.

Gallego, E., Roca, F. J., Perales, J. F., & Guardino, X., (2010). Comparative study of the adsorption performance of a multi-sorbent bed (Carbotrap, Carbopack X, Carboxen 569) and Tenax TA for the analysis of volatile organic compounds (VOCs). *Talanta.*, *81*, 916–924.

Giddings, J. C., Seager, S. L., Stucki, L. R., & Stewart, G. H., (1960). Plate height in gas chromatography. *Anal. Chem.*, *32*, 867–870.

Griffith, K. S., & Gellene, G. I., (1993). A simple method for estimating effective ion source residence time. *J. Am. Soc. Mass Spectrom.*, *4*, 787–791.

Grob, K., (2008). *Split and Splitless Injection for Quantitative Gas Chromatography, 4th edn.* Wiley-VCH: Federal republic of Germany.

Grob, R. L., & Barry, E. F., (2004). *Modern Practice of Gas Chromatography, 4th edn.* Wiley and Sons: New York.

Guilhaus, M., (1995). Principle and instrumentation in time-of-flight mass spectrometry. *J. Mass Spectrom.*, *30*, 1519–1532.

Halket, J. M., Przyborowski, A., Stein, S. E., Mallard, W. G., Down, S., & Chalmers, R. A., (1999). Deconvolution gas chromatography/mass spectrometry of urinary organic acids-potential for pattern recognition and automated identification of metabolic disorders. *Rapid Commun. Mass Spectrom.*, *13*, 279–284.

Handley, A.J., & Adlard, E.R., (2001). *Gas Chromatographic Techniques and Applications, 1ˢᵗ edn.*, Sheffield Academics: London, Book Series: Sheffield Analytical Chemistry (Book 5), ISBN-10: 0849305217, ISBN-13: 978-0849305214, pp. 320.

Horai, H., Arita, M., Kanaya, S., Nihei, Y., Ikeda, T., Suwa, K., Ojima, Y., Tanaka, K., Tanaka, S., Aoshima, K., Oda, Y., Kakazu, Y., Kusano, M., Tohge, T., Matsuda, F., Sawada, Y., Hirai, M. Y., Nakanishi, H., Ikeda, K., Akimoto, N., Maoka, T., Takahashi, H., Ara, T., Sakurai, N., Suzuki, H., Shibata, D., Neumann, S., Iida, T., Tanaka, K., Funatsu, K., Matsuura, F., Soga, T., Taguchi, R., Saito, K., & Nishioka, T., (2010). Mass Bank: a public repository for sharing mass spectral data for life science. *J. Mass Spectrom.*, *45*, 703–714.

Hu, Q., Noll, R. J., Li, H., Makarov, A., Hardman, M., & Cook, G. R., (2005). The Orbitrap: a new mass spectrometer. *J. Mass Spectrom.*, *40*, 430–443.

IUPAC, (2017). *Compendium of Chemical Terminology*, 2nd ed. (the "Gold Book"). Compiled by McNaught, A. D., & Wilkinson, A. Blackwell Scientific Publications, Oxford (1997). XML on-line corrected version: http://goldbook.iupac.org (2006) created by Nic, M., Jirat, J., & Kosata, B., updates compiled by Jenkins, A. ISBN 0–9678550–9–8. https://doi.org/10.1351/gold book (accessed May 07, 2017).

James, A. T., Martin, A. J. P., & Howard Smith, G., (1952). Gas-liquid partition chromatography, the separation and micro-estimation of ammonia and methylamines. *Biochem. J.*, *52*, 238–242.

Johnson, J. V., Yost, R. A., Kelley, P. E., & Bradford, D. C., (1990). Tandem-in-space and tandem-in-time mass spectrometry: triple quadrupoles and quadrupoles ion traps. *Anal. Chem.*, *62*, 2162–2172.

Jonsson, P., Johansson, A. I., Gullberg, J., Trygg, J., Jiye, A., Grunge, B., Marklund, S., Sjöström Antti, H., & Moritz, T., (2005). High-Though data analysis for detecting and identifying differences between samples in GC-MS based metabolomics analyses. *Anal. Chem.*, *77*, 5635–5642.

Kokkonen, J., Leinonen, A., Tuominen, J., & Seppälä, T., (1999). Comparison of sensitivity between gas chromatography low resolution mass spectrometry and gas chromatography high resolution mass spectrometry for determining methandienone metabolites in urine. *J. Chromatogr. B. Biomed. Sci. Appl.*, *734*, 179–189.

Luong, J., Gras, R., Mustacich, R., & Cortes, H., (2006). Low thermal mass gas chromatography: principle and applications. *J. Chromatogr. Sci.*, *44*, 253–261.

Makarov, A., Denisov, E., Kholomeev, A., Balschun, W., Lange, O., Strupat, K., & Horning, S., (2006). Performance evaluation of a hybrid linear ion trap/orbitrap mass spectrometer. *Anal. Chem.*, *78*, 2113–2120.

Mamyrin, B. A., (2001). Time-of-flight mass spectrometry (concepts, achievements, and prospects). *Int. J. Mass Spectrom.*, *206*, 251–266.

Märk, T. D., & Dunn, G. H., (1985). *Electron Impact Ionization, 1st edn.* Springer-Verlag: Wien, ISBN-10: 3709140307, pp. 384.

Nordström, A., & Lewenson, R., (2010). Metabolomics: Moving to the Clinics. *J. Neuroimmune Pharmacol.*, *5*, 4–17.

Perry, R. H., Cooks, R. G., & Noll, R. J., (2008). Orbitrap mass spectrometry: instrumentation, ion motion and applications. *Mass Spectrom. Rev.*, *27*, 661–699.

Peterson, A. C., McAlister, G. C., Quarmby, S. T., Griep-Raming, J., & Coon, J. J., (2010). Development and characterization of a GC-Enabled QLT-Orbitrap for high resolution and mass accuracy GC-MS. *Anal. Chem.*, *82*, 8618–8628.

Restek, (2016), Guide to GC column selection and optimizing separations, www.restek.fr/pdfs/GNBR1724-UNV. pdf (accessed Jun 21, 2016).

Robroeks, C. M. H. H. T., Van Berkel, J. J. B. N., Dallinga, J. W., Jöbsis, Q., Zimmermann, L. J. I., Hendriks, H. J. E., Wouters, M. F. M., Van der Grinten, P. M., Van de Kant, K. D. G., Van Schooten, F. J., & Dompeling, E., (2010). Metabolomics of volatile organic compounds in cystic fibrosis patients and control. *Pediatr. Res.*, *68*, 75–80.

Rudnicka, J., Kowalkowski, T., Ligor, T., & Buszewski, B., (2011). Determination of volatile organic compounds as biomarkers of lung cancer by SPME-GC-TOF/MS and chemometrics. *J. Chromatogr. B.*, *879*, 3360–3366.

Skogerson, K., Wohlgemuth, G., Barupal, D. K., & Fiehn, O., (2011). The volatile compound binbase mass spectral database. *BMC Bioinformatics*, *12*, 321.

Sloan, K. M., Mustacich, R. V., & Eckenrode, B. A., (2001). Development and Evaluation of a low thermal mass gas chromatograph for rapid forensic GC-MS analysis. *Field Anal. Tech.*, *5*, 288–301.

Smith, C. A., Mallie, G. O., Want, E. J., Qin, C., Trauger, S. A., Brandon, T. R., Darlene, E., Abagyan, R., & Siuzdak, G., (2005). Metlin: A Metabolite Mass Spectral Database. *Ther. Drug Monit.*, *27*, 747–751.

Snow, N. H., & Slack, G. C., (2002). Head space analysis in modern gas chromatography. *Trends Anal. Chem.*, *21*, 608–617.

Stein, S. E., (1999). An integrated method for spectrum extraction and compound identification from gas chromatography/mass spectrometry data. *J. Am. Soc. Mass Spectrom.*, *10*, 770–781.

Styczynski, M. P., Moxley, J. F., Tong, L. V., Walther, J. L., Jensen, K. L., & Stephanopoulos, G. N., (2007). Systematic identification of conserved metabolite in GC-MS data for Metabolomics and biomarkers discovery. *Anal. Chem.*, *79*, 966–973.

Synge Richard, L. M., (2017). Facts in Nobelprize.org. http://www.nobelprize.org/nobel_prizes/chemistry/laureates/1952/synge-facts.html (accessed May 06, 2017).

Todd, J. F. J., (1995). Recommendation for nomenclature and symbolism for mass spectroscopy including an appendix of term used in vacuum technology. *Int. J. Mass Spectrum. Ion Process*, *142*, 211–240.

Tsugawa, H., Tsujimoto, Y., Arita, M., Bamba, T., & Fukusaki, E., (2010). GC-MS based metabolomics development of data mining system for metabolite identification by using soft independent modeling of class analogy (SIMCA). *BMC Bioinformatics*, *12*, 131.

Tsugawa, H., Tsujimoto, Y., Sugitate, K., Sakui, N., Nishiumi, S., Bamba, T., & Fukusaki, E., (2014). Highly sensitive and selective analysis of widely targeted metabolomics using gas chromatography/triple-quadrupole mass spectrometry. *J. Biosci. Bioeng.*, *117*, 122–128.

Van Deemter, J. J., Zuiderweg, F. J., & Klinkenberg, A., (1956). Longitudinal diffusion and resistance to mass transfer as cause of nonideality in chromatography. *Chem. Eng. Sci.*, *5*, 271–289.

Wishart, D. S., Jewison, T., Guo, A. C., Wilson, M., Knox, C., Lui, Y., Djoumbou, Y., Mandal, R., Aziat, F., Dong, E., Bouatra, S., Sinelnikov, I., Arndt, D., Xia, J., Liu, P., Yallou, F., Bjorndahl, T., Perez-Pineiro, R., Eisner, R., Allen, F., Neveu, V., Greiner, R., & Scalbert, A., (2012). HMDB 3.0. The Human Metabolome database in 2013. *Nucleic Acids Res.*, *41*, 801–807.

Yost, R. A., & Enke, C. G., (1979). High efficiency collision-induced dissociation in a RF only quadrupole. *Int. J. Mass Spectrom. Ion Phys.*, *30*, 127–136.

Yu, S., Crawford, E., Tice, J., Musselman, B., & Wu, J. T., (2008). Bioanalysis with sample clean up or chromatography: The Evaluation and initial implementation of direct analysis in real time ionization mass spectrometry for the quantification of drugs in biological matrixes. *Anal. Chem.*, *81*, 193–202.

PART II
Biomedical Diagnosis Applications

CHAPTER 3

ANALYSIS OF VOLATILE ORGANIC COMPOUNDS FOR CANCER DIAGNOSIS

ABIGAIL V. RUTTER[1] and JOSEP SULÉ-SUSO[1,2]

[1]Institute for Science and Technology in Medicine, Keele University, Guy Hilton Research Centre, Thornburrow Drive, Stoke on Trent ST4 7QB, UK

[2]Oncology Department, Royal Stoke University Hospital, University Hospitals of North Midlands (UHNM), Newcastle Rd, Stoke on Trent, Staffordshire ST4 6QG, UK, E-mail: josep.sulesuso@uhns.nhs.uk

ABSTRACT

For centuries, the smell of breath has been associated with certain diseases. This has attracted worldwide interest on using breath analysis for disease diagnosis. The possibility of using breath analysis to improve the management of cancer has led to the identification of volatile organic compounds (VOCs) linked to cancer. It could be assumed that the proof of concept has been more or less established. However, despite the vast amount of work on this subject, the study of breath has not made it into clinical practice yet. The reasons behind this are multiple and range from identifying the best sampling method and the origin of these VOCs, all the way to instrumentation and data analysis standardization. Further work on these issues will need to be followed with multicenter studies to confirm that breath analysis is not only a potential approach to follow in clinical practice but rather, a further robust methodology made available to clinicians in order to improve the care of patients with cancer.

3.1   INTRODUCTION

Primum non nocere (first, do no harm) is a fundamental principle in medicine, taught in medical schools all over the world reminding health professionals to consider the possible harm that any intervention could cause to patients. The quest to develop interventions for the management of disease that would cause as little as possible harm to patients goes back to ancient Greece and is still an important issue to be taken into account in medical research today. Thus, it does not come as a surprise that something not invasive, such as using breath for the management of disease, draws researchers' attention worldwide.

The study of the smell of breath in the management of disease goes back to ancient Greece when Hippocrates described *fetor oris* and *fetor hepaticus*. From then on over the following centuries, people were aware through observation that some breath smells were related to certain diseases. However, it was not until Pauling's work, identifying over 200 different volatile organic compounds (VOCs) in human exhaled air and urine headspace by gas chromatography (GC), which was a major break-through in this field was achieved (Pauling et al., 1971). It was, therefore, obvious that from this point forward further work was needed to assess whether the study of VOCs from breath and biological fluids could be used as a non-invasive tool for the diagnosis and management of diseases in a more scientific way. Well-known examples of this are the ($^{13/14}$C) urea breath test (UBT) used in the diagnosis of *Helicobacter pylori* infection (Gisbert and Pajares, 2004; Logan, 1998; Romagnuolo et al., 2002) and the nitric oxide (NO) breath test in the diagnosis of airway inflammation (Miekisch et al., 2004; Rosias et al., 2004; Stick, 2002). More importantly, the parallel development of new technologies such as the Selected Ion Flow Tube Mass Spectrometry, Proton Transfer Reaction Mass Spectrometry, and Ion Mobility Spectrometry amongst others (reviewed by Sun et al., 2016) has provided new tools to explore the potential of VOC analysis in the management of disease and its application in clinical practice. Further-more, VOCs can now be measured in concentrations in the range of parts per million (ppm) to parts per trillion (ppt) by volume (Miekisch et al., 2004; Mukhopadhyay, 2004; Schubert et al., 2004). However, while breath samples might contain around 3000 different VOCs, the significance for most of them is unknown (Phillips et al., 2013). Therefore, further work is needed to understand the metabolism and production of VOCs, which should help towards the implementation of volatiles' analysis in clinical practice for disease management.

A disease that researchers have promptly chosen for the study of VOCs is cancer. Cancer is a major health and socio-economic problem worldwide. According to the World Health Organization (WHO), cancer is one of the leading causes of morbidity and mortality worldwide with approximately 14 million new cases and 8.8 million cancer-related deaths in 2015. More worrying, the number of new cases is expected to rise by about 70% over the next 2 decades (WHO, 2016). Cancer is associated in many cases with difficulties in developing successful and robust screening tools in order to achieve an early diagnosis, and there is also a lack of treatments (especially patient-tailored therapies see below) that would provide not only long-lasting survival but also reduced cancer death rates; and is associated in too many cases with a poor prognosis. The accumulation of these problems makes this disease an ideal area of research that can and should be further developed, establishing methodologies that utilize the identification of VOCs as biomarkers of screening/diagnosis, the assessment of treatment response and follow-up, amongst other. However, in the quest of VOCs as potential biomarkers for cancer, the criteria for an "ideal" biomarker should be clarified. An ideal biomarker/s should:

1. be sensitive and specific for the diagnosis of the disease process.
2. reflect, or be a very clear surrogate, of the pathophysiologic mechanism.
3. be stable and only vary with events known to relate to disease progression.
4. predict early-stage disease development.
5. predict disease progression.
6. be responsive to interventions known to be effective.
7. to be used in simple and low-cost point-of-care diagnostic devices for low-income countries (Stockley, 2007).

Although it is evident that it would be ideal to have a VOC or combination of VOCs as biomarker/s for cancer management fulfilling all these criteria, this would be daunting and challenging to achieve, especially in cancer management. In fact, while for some diseases such as *P. aeruginosa* infection a single VOC might suffice (Smith and Španěl, 2015), in the case of cancer it is believed that a combination of VOCs rather than a single

VOC will be needed to screen/diagnose the disease and/or improve the management of cancer patients. Furthermore, there are several steps to be taken before a biomarker or combination of biomarkers can be used in clinical practice. The Early Detection Research Network (EDRN) has identified five phases from the discovery of a biomarker and its clinical implementation:

1. preclinical exploratory studies;
2. clinical assay development for clinical disease;
3. retrospective longitudinal repository studies;
4. prospective screening studies; and
5. case-control studies (reviewed by Hensing and Salgia, 2013).

In the case of cancer, the changes in cells leading to the appearance and further development of the disease are linked to changes in cell metabolism that differ from the metabolism of non-malignant cells. Clear examples are metabolic pathways linked to glycolysis, apoptosis, loss of tumor suppressor genes, and angiogenesis. These pathways can be modified, inhibited and activated or over-activated (Jones and Thompson, 2009). Thus, it could be hypothesized that changes in some of these pathways may alter the production of VOCs in the body. However, it is not clear yet whether and to what extent VOCs are associated with cancer cell metabolism or whether the microenvironment around cancer cells could be the source of VOCs that could be used as biomarkers for cancer management. Two metabolical pathways that have been extensively studied in cancer are oxidative stress and the Warburg effect. Oxidative stress is the overall balance between the production of free radicals and the ability of the body to counteract their harmful effects through neutralization by antioxidants. It is also one of the main sources of developing cancer via the overproduction of reactive oxygen species (ROS) and nitrogen species resulting in mutations (Toyokuni, 2008). ROS are molecules or ions such as peroxide (H_2O_2), hydroxyl radical (OH) and superoxide radical (O_2) with an unpaired electron in the outer shell, which are constantly produced in the mitochondria as part of the cellular respiration process (Ghezzi et al., 2016). During oxidative stress, ROS are excreted from the mitochondria in the cell or from the peroxidate polyunsaturated fatty acids in the cell membranes generating volatile alkanes that are emitted in the breath (Hakim et al.,

2012). However, ROS have a complex metabolism and are generated by different enzymes at diverse sites and with different timing. Furthermore, different diseases including inflammatory conditions, infections and exogenous sources (pollution, cigarette smoke, radiation) are associated with ROS production (reviewed by Ghezzi et al., 2016). Therefore, caution should be taken when studying VOCs linked with oxidative stress in cancer management.

In the metabolism of glucose by mammalian cells, energy is harnessed in the form of ATP through the oxidation of its carbon bonds, producing lactate as the end product or, upon full oxidation of glucose through oxidative phosphorylation in the mitochondria, CO_2 (Liberti and Locasale, 2016). However, in tumor cells, the rate of glucose metabolism increases and lactate is produced even in the presence of oxygen and fully functioning mitochondria (aerobic glycolysis) (Liberti and Locasale, 2016). The change in the metabolism of glucose in cancer cells favoring glycolysis to oxidative phosphorylation even in the presence of oxygen is known as Warburg effect (Warburg, 1956) and is characterized by an increased glucose consumption and lactate production. The increased levels of lactate production lead to the cancer cell microenvironment and tissue to become acidic. This acidic environment permits breakage of the basement membrane and allows accessibility of cancer cells to blood vessels, which could lead to the development of metastases through blood (Gatenby and Gillies, 2004). Therefore, it could be hypothesized that these changes in cancer cells metabolism, when compared to non-malignant cells, could lead to the release of VOCs in breath linked to cancer.

The possibilities of measuring VOCs in the breath of patients with cancer would obviously lead to lung cancer being one of the main targets. However, other tumors could also be identified through breath analysis. Examples are lung metastasis from tumors originating outside the lungs, which could also produce their own VOC pattern. On the other hand, solid tumors in other organs and leukemia cells present in blood could also produce volatiles that could be released in the breath of affected individuals thanks to the exchange of volatiles through the blood-gas interface in lung alveoli (Sapoval et al., 2002). It has been estimated that the mean number of alveoli in a healthy lung is around 480×10^6 (Ochs et al., 2004) providing a surface area of around 70 m^2 (Notter, 2000). This indicates that breath analysis could be a potential tool in the management of different types of tumors. Furthermore, the advantages of breath testing are, amongst others:

1. it is non-invasive;
2. breath samples closely reflect the blood concentrations of some biological substances;
3. breath is a much less complicated mixture than serum or urine;
4. it provides direct information on respiratory function;
5. its analysis can dynamically and in real-time monitor volatiles (Cao and Duan, 2006).

However, breath testing also has certain limitations like:

1. lack of standardization of analytical methods and a wide variation in results obtained in different studies;
2. the high water content of breath samples, which may affect pre-concentration, separation, and detection of single compounds;
3. several instruments for breath analysis are expensive;
4. a lack of clearly established links between breath substances and disease.

As an example, ethane is produced as a result of lipid peroxidation, but ethane in breath could also be derived from environmental sources and bacteria in the gastrointestinal tract (Cao and Duan, 2006). However, despite these limitations, several studies have shown a possible relation between VOCs present in breath and different types of cancer (Haick et al., 2014; Krilaviciute et al., 2015; Lourenço and Turner, 2014; Schmidt and Podmore, 2015; Sun et al., 2016).

There have already been several recent reviews published on different techniques, methodologies and sampling approaches used for VOCs' analysis and cancer, and reviews on the VOCs that could have a possible clinical application in the management of cancer (Adiguzel and Kulah, 2015; Dent et al., 2013; Haick et al., 2014; Krilaviciute et al., 2015; Sun et al., 2016). The aim of this chapter is not to iterate some of the already published work on the different techniques available to study VOCs that might be linked to cancer but rather, to discuss some of the possible difficulties that need tackling before the analysis of VOCs could be used in routine clinical practice for cancer management.

3.2 VOCS' RESEARCH IN CANCER

3.2.1 IN VITRO STUDIES

It is not yet clear whether some of the VOCs linked to cancer are indeed produced by cancer cells or by the environment surrounding cancer cells which contain, amongst other, the extracellular matrix and different cell types such as tissue supporting cells and cells participating in the immune response against cancer. Furthermore, it has to be acknowledged that cancer is made up of different cancer cells with different proliferative characteristics and different grades of resistance to external agents such as chemotherapy, radiotherapy or targeted therapies. Furthermore, for a given time, a tumor will be made up of proliferating cells, quiescent cells (alive but not proliferating) and dying/dead cells in different combinations. These tumoral characteristics might alter the VOCs pattern for each individual tumor. Furthermore, the immune response that the human body displays against cancer might be different for each individual tumor and patient. This tumor microenvironment will thus affect the release of VOCs measured in, as an example, breath.

In vitro studies offer the possibility of tackling some of these issues by allowing the growth of cancer cells alone or in cell mixtures. Several groups have studied the release of VOCs by cancer cells in different culture conditions and prior to and after the addition of drugs (Brunner et al., 2010; Filipiak et al., 2008, 2010; Rutter et al., 2013; Sponring et al., 2009, 2010; Sulé-Suso, 2009). However, most of these studies have used whole populations of cancer cells without taking into account intra-tumor cell variability. In fact, a study found differences in VOCs' levels between not only different cell lines of the same type of cancer but also between cells within the same cell line (Schmidt and Podmore, 2015). It is, therefore, important to assess how the VOCs' profile varies between tumor cells within a given tumor cell population. A way to answer this could be by using cell cloning. The possibility of obtaining a cancer cell clone from a single cancer cell provides the advantage of identifying possible differences in VOCs' pattern from different cells within a given cancer cell population. Figure 3.1 shows a schematic diagram of a methodology for lung cancer cell cloning (Rutter et al., 2014). Briefly, cancer cells are obtained from a culture flask and seeded at 0.5 cells per well in 200 µL of fresh medium in 96 flat-bottomed well plates (example and for simplicity, for 100 wells, 50 cells are placed in 20 mL of medium and distributed evenly in these

100 wells). Clones are allowed to grow for 2 weeks. Clones are regularly inspected so those wells with more than 1 clone can be discarded. This experimental setup can yield between 25 and 30 clones from a total of five 96 well plates for the lung cancer cell line CALU-1 (Rutter et al., 2014). The methodology described here can also help to develop clones that are resistant, semi-resistant/semi-sensitive or sensitive to a chemotherapy drug or combination of drugs or other external agents (Rutter et al., 2014). The measurement of VOCs released from these cancer cell clones prior to and after the addition of drugs could be used to assess whether there might be a link between the VOCs' profile of cancer cells and the effects of drugs on cells in more uniform cancer cell populations. The clinical translation would be to assess through breath tumor response to treatment (see below).

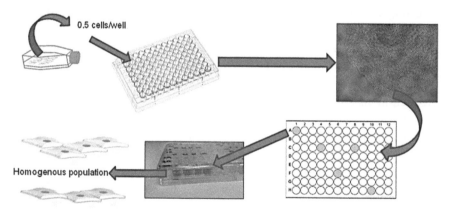

FIGURE 3.1 **(See color insert.)** Schematic protocol to isolate cell clones from cancer cell lines.

In vitro studies can also be used to assess how the extracellular matrix and other non-malignant cells can affect the release of VOCs in the tumor microenvironment. The growth of cancer cells in 3D models such as collagen gels (Rutter et al., 2013) are a better approximation to the physiological situation of cells growing *in vivo* when compared to 2D models. This will also allow the addition of extracellular matrix components and non-malignant cells in a controlled way, which should help to clarify the origin of certain VOCs better.

An important issue in the diagnosis of cancer through the measurement of volatiles is to be able to diagnose cancer in early stages. It is widely known that prognosis is linked to tumor stage, the earlier the tumor is detected and treated the better the prognosis. Therefore, it is important that VOCs' analysis being able to detect tumors prior to patients presenting with symptoms. However, it is not clear yet for a given tumor, the minimum number of cells needed to release VOCs in a concentration sufficient to be detected with presently available techniques. *In vitro* studies using 3D models could represent an excellent opportunity to understand this better. Therefore, while it is important to carry out the analysis of VOCs *in vivo*, the *in vitro* studies need to be run in parallel to understand better the results obtained in clinical studies. However, *in vitro* studies do not guarantee that the measured VOCs are of endogenous origin. They may not be produced by cancer cells themselves and may instead come from other sources such as culture vessels, extraction devices, and the sampling environment (Kwak et al., 2013; Zimmermann et al., 2007). Also, it is possible that the VOCs' profile from *in vitro* culture cells could also change with cell passage (Sponring et al., 2009). On the other hand, it would still be difficult to address the study of possible VOCs released by the so-called cancer stem cells (CSC) for *in vitro* models due to the difficulty in isolating and expanding these cells from certain tumors. The existence of CSC was first proposed by Hamburger and Salmon (1977). It is believed that some cancers originate from cells with stem cell properties including self-renewal, and multilineage differentiation capacity (Jorgensen, 2009; McDonald et al., 2009). The malignant transformation of an adult stem cell leads to clonal expansion of these cells, which fully transform into cancer. However, despite these issues, several studies have shown differences between VOCs released by cancer cells and non-malignant cells *in vitro* (reviewed by Schmidt and Podmore, 2015). The next step is to correlate the VOCs measured *in vitro* for a given tumor with those measured in patients' with the same type of tumor. While the release of VOCs by lung cancer cells *in vitro* and by the breath of patients with lung cancer have been compared (Chen et al., 2009; Kalluri et al., 2014), the differences found between them might not be sufficient at the moment for *in vitro* cultures to be a good model for the VOCs present in exhaled breath (Kalluri et al., 2014). An example is hypoxia, which affects cancer cell metabolism and might also affect the VOCs released by cancer cells. As hypoxic culture conditions might resemble more the cancer cell growth *in vivo*, it has been suggested that the measurement of VOCs from *in vitro* growing cancer cells should include cell growth in hypoxic conditions (Kalluri et al., 2014).

3.2.2   EX VIVO STUDIES

Ex vivo studies have been carried out to analyze the VOCs released by body fluids such as blood, saliva, urine and feces and even from tissue samples from patients with cancer (reviewed by de Lacy Costello et al., 2014). The main aim has been to identify VOCs released by body fluids, which could be used in cancer management. On the other hand, several studies have also been carried out to better understand the relationship between VOCs present in blood and breath. In the case of blood, it has been described that the number of volatiles identified in blood are relatively few compared to those identified in breath (Mochalski et al., 2013). However, the VOCs' profile in blood will also change depending on whether the individual has other non-malignant diseases including infection. The VOCs released from blood from patients with different types of cancer have been studied (Deng et al., 2004a, 2004b; Xue et al., 2008). As an example, it would seem that the VOCs' profile released from blood from lung cancer patients can be used to identify lung cancer as some of these VOCs were also found in the breath of patients with lung cancer (Deng et al., 2004a). However, some of the problems of measuring VOCs from blood samples are its invasiveness, which in the case of children could be rather distressing for both children and their parents and relatives, and also the way samples are transported and handled after collection as both temperature and pH can alter the VOCs measured (Kouremenos et al., 2012; Manolis, 2007). Another body fluid that has been used to study the relationship between its VOCs' profile and cancer is urine. The advantage of using urine instead of blood is its non-invasive obtention. Interestingly, urine has been used to study the relationship between its VOCs' profile and not only bladder cancer (Jobu et al., 2012) but also other types of cancer (Guadagni et al., 2011; Huang et al., 2013; Silva et al., 2012).

Another area of cancer research, which is gaining interest, is the detection of circulating tumor cells (CTC) in blood. CTCs originate from primary or metastatic solid tumor and are present in the circulation at extremely low levels (Alix-Panabieres and Pantel, 2013; Saad and Pantel, 2012; Zhang and Ge, 2013). However, technical improvements for the isolation and enrichment of CTCs have led to use this concept as a tool to assess tumor diagnosis, prognosis, tumor recurrence, and presence of metastases (Huang et al., 2015; Katoh et al., 2015; Tang et al., 2013; Tinhofer et al., 2014; Unesono et al., 2013; Wang et al., 2014). However, identifying these rare CTCs is still a challenge before a clinical application can be developed (Zhang et al., 2015). It is not possible at present to say whether the presence of these rare CTCs in blood could lead to the production of VOCs' profiles from blood samples that could indicate the presence of these cells in the blood.

Cancer tissues in an *ex vivo* set up have also been used to study their VOCs' profile (Buszewski et al., 2012; Filipiak et al., 2014). The advantage of using cancer tissue is that VOCs released by cancer cells and tissues can be measured reducing contamination from other metabolical processes and/or diseases present in the patient. It has been shown that the 27 VOCs were detected from cancer tissues from lung cancer patients, indicating that the amount of VOCs to be used as biomarkers for cancer management can be reduced to a more manageable number. In fact, mainly alcohols, aldehydes, ketones, and aromatic and aliphatic hydrocarbons were detected in the headspace of lung cancer tissue and breath from patients with lung cancer (Buszewski et al., 2012).

The detection of VOCs from body fluids and tissue (malignant or not) has the potential to help in the management of cancer. However, there are several issues to be taken into account. In the case of blood, the techniques will rely on an invasive procedure which, although it could be acceptable in the adult population, it would be less desirable in children. However, thanks to the blood-gas interface in lung alveoli, VOCs present in blood could also be detected in breath. Therefore, further studies are needed to understand better the potential of measuring VOCs released from blood samples when compared to VOCs present in breath for cancer management. On the other hand, urine can be obtained in a non-invasive way and, furthermore, it could be in contact with tumors of the urinary system. It could be hypothesized that measuring VOCs released from urine would be better than measuring VOCs present in breath for the diagnosis, assessment of tumor response and/or identifying tumor recurrence for urinary tract tumors. While this would seem more plausible for tumors of the urethra, bladder, and ureter thanks to the contact of urine with the tumor itself, this might be less plausible for tumors of the kidney and prostate, which might not be in direct contact with the excreted urine.

3.2.3 *IN VIVO STUDIES*

There have been several studies measuring the VOCs in the breath of patients with different types of cancer including lung, breast, gastric and colorectal (Altomare et al., 2013; Amal et al., 2016; Barash et al., 2015; Corradi et al., 2015; Kumar et al., 2013; Peled et al., 2012; Phillips et al., 2003b; Wang et al., 2014; Wehinger, 2007). These studies have been able to differentiate patients with cancer and control cases without cancer. It seems that the proof of concept that the detection of VOCs in breath has the potential to help clinicians in the management of cancer has already been demonstrated. It could be argued that for cancer management, the identification of a given VOC profile should suffice even if the biochemical/ metabolical origin of this VOC profile is unknown. At present, the origin of some volatile biomarkers for some types of cancer remains speculative (Phillips et al., 2010). Also, more knowledge is needed on the paths followed by VOCs once they are released by tumor cells and enter the body fluids before they are finally expelled through breath (Hakim et al., 2012).

There is still a lot of work and effort to be put in place to tackle different issues before breath analysis makes it into a clinical set up as part of the armamentarium available to clinicians to screen, diagnose, treat and carry out the follow-up of patients with cancer. Patients' characteristics have to be also taken into account. These include, amongst other, age, gender, ethnicity, smoking, food intake, other medical conditions and treatments patients might be receiving. They could affect the VOCs' profile in breath (Di Francesco et al., 2005; Kischkel et al., 2010). It has been reported that there could be a difference between certain VOCs measured in breath depending on the age of the subject studied (Phillips et al., 2000; Smith et al., 2010). However, other studies showed different results when taking age as a possible reason behind changes in VOCs' profile in breath (Dragonieri et al., 2007; Mazzone et al., 2007; Wehinger, 2007). Regarding gender, several studies have shown that gender has no effect on the VOCs profile (Mazzone et al., Peng et al., 2010; 2007; Wehniger, 2007). However, it has been reported that isoprene was increased in the breath of male subjects (Lechner et al., 2006). One of the symptoms in cancer patients is weight loss. Certain compounds such as the formation of ketone bodies including acetone are related to weight loss but are also seen in cancer progression due to an increase in the rate of fatty acid oxidation (Murray et al., 2006). Ethnicity might also need to be taken into account when studying the VOC profile in breath if this technique is to be implemented worldwide. It has been shown that populations from two distinct geographical regions (China and Latvia) have separate VOCs' profile origin (Amal et al., 2013). Oral hygiene is also an important factor to be taken into account when measuring VOCs from breath. Several compounds might be produced by anaerobic bacteria in oral cavity and nose (Khalid et al., 2013; Smith et al., 2013; Španěl et al., 2006). Furthermore, VOCs can be produced by bacteria in the gastrointestinal tract and excreted by the lungs (Amann et al., 2010), an obvious example being *Helicobacter pylori* present in the stomach (Ulanowska et al., 2011). However, those volatiles produced in the gut and entering the blood stream might not be detected in breath due to conversion in the liver (de Lacy Costello et al., 2014). On the other hand, VOCs transformation can occur not only in the lungs but also via enzymes in the nose (Ding and Kaminsky, 2003).

The VOCs released in breath can be affected to different extent by lung diseases (Chronic Obstructive Pulmonary Disease (COPD), asthma, emphysema, etc.) as well as by metabolical disorders (diabetes, phenylketonuria, metabolic syndrome, sodium metabolism disorders, calcium metabolism disorders, hyper- and hypocalcemia, potassium metabolism disorders, hyper- and hypokalemia, phosphate metabolism disorders, magnesium metabolism disorders, and acid–base metabolism disorders) (Cao and Duan, 2006). Smoking will also affect the VOCs' profile in breath with several compounds found in higher concentration in smokers when compared to non-smokers (Kalluri et al., 2014). Similarly, food intake will also affect the VOCs present in breath (Smith and Španěl, 2005). Even a patient's position during sampling (Beauchamp and Pleil, 2013) or exertion of an effort can affect the release of certain VOCs such as isoprene (King et al., 2009, 2010).

All these factors linked to patients affecting VOCs' profile in breath need to be taken into account and better understood before a full clinical application of breath analysis in the management of cancer is fully developed. Many of these issues will have to be answered with multicenter studies including big numbers of patients with robust statistical and data analysis approaches.

3.3   VOC'S IN CANCER MANAGEMENT

3.3.1   SCREENING/DIAGNOSIS

The prognosis of cancer depends on several factors, which could vary between the different types of cancer. However, one factor common to all cancers is the stage of the disease at diagnosis. For a given tumor, the prognosis worsens with higher the stage at diagnosis. In other words, the prognosis worsens the more advanced the disease is at the initial diagnosis. Diagnosing the disease at earlier stages is one of the reasons behind the work carried out in cancer screening. However, to be able to diagnose cancer before patients present with symptoms through screening will not suffice. There are several issues that need to be taken into account when developing screening programs. First, the disease itself should be sufficiently burdensome to the population that a screening program is warranted (Phillips, 2003a). Once the disease is identified, the diagnosis should be early enough that treatments are effective and will improve outcomes (Hensing and Salgia, 2013). Furthermore, the screening technique should be inexpensive enough to be cost-effective, providing results that are reproducible and with minimal false-positive and false-negative results (Hensing and Salgia, 2013; Phillips, 2003a). On the other hand, screening programs have to take into account the possible bias that can be introduced. The different types of bias in screening are listed in Box 1. As an example, screening for a given tumor might lead to the disease being diagnosed earlier (before patients present with symptoms) so treatment can be started earlier. This might increase survival but not improve death rates from the type of cancer screened. In other words, the time of death for these patients would remain unchanged. The increased survival is simply due to the disease being diagnosed earlier (lead-time bias). Lung cancer is an excellent example where screening is not being carried out routinely like in other types of tumors such as breast cancer (screened with mammography) as screening for lung cancer has not translated into improved death rates.

Once a tumor has been diagnosed, it is important in some cases to identify the subtype of cancer, as the management could be different for different cancer subtypes. An example is lung cancer. Table 3.1 shows a modified WHO classification of the different types of lung tumors. While non-small cell lung cancer can be treated with surgery, chemotherapy, radiotherapy, targeted therapies or a combination of these, surgery is not indicated for small cell lung cancer unless for diagnostic purposes. It is difficult to say whether breath analysis will be able in the future to subclassify the different subtypes of cancer or just a few subtypes. This is also important when dealing with metastatic tumors from an unknown origin. Breath analysis might be a valuable tool for clinicians in the future to help them to identify the source of some metastatic tumors so a quicker and more patient-tailored treatment can be established.

TABLE 3.1 Modified WHO Classification of Malignant Lung Tumors

Histologic Type and Subtypes	
Malignant epithelial tumors	**Benign epithelial tumors**
Squamous cell carcinoma	Papillomas
Adenocarcinoma	Adenomas
Large cell carcinoma	
Small cell carcinoma	**Lymphoproliferative tumors**
Adenosquamous carcinoma	
Sarcomatoid carcinoma	**Miscellaneous tumors**
Carcinoid tumor	Hamartoma
Salivary gland tumor	Sclerosing hemangioma
Pre-invasive lesions	Clear cell tumor
Mesenchymal tumors	Germ cell tumors
	Intrapulmonary thymoma
	Melanoma
	Metastatic tumors

3.3.2 TREATMENT

The choice of chemotherapy agents and drug combinations in the treatment of cancer is based on multicenter studies resulting in the identification of anticancer drugs and their combinations giving the best results for individual types of cancer. However, it is not clear yet how the best drug combination can be identified for each individual patient with cancer. There are several chemotherapy sensitivity/resistance assays such as gene expression profiling (Chang et al., 2003; Robert et al., 2004), *in vitro* clonogenic and proliferation assays, cell metabolic activity assays, molecular assays, *in vivo* tumor growth and survival assays, and *in vivo* imaging assays amongst other (Blumenthal and Goldenberg, 2007; Cree, 2009). However, it is not clear yet whether these assays can improve the outcome of patients with cancer and, so far, the clinical application has proved difficult. On the other hand, in order to assess tumor response to treatment, clinicians use, amongst other, physical examination, blood tests (looking for the presence and concentration of tumor markers in blood) and imaging techniques (chest X-rays, CT scan, MRI scan, PET scan, bone scan, …). Some of these could be invasive, costly and give radiation.

Therefore, it would be ideal to develop techniques that are cheaper, non-invasive and radiation free.

Breath analysis has again the potential to be such a technique. *In vitro* studies, as described above will be needed to assess whether VOCs analysis can identify volatile markers of tumor sensitivity/resistance to a drug or combination of drugs. Furthermore, breath analysis could be used to evaluate in a fast, non-invasive way tumor response to treatment at different time points. In fact, studying VOCs' profile prior to and after a given treatment in the same patient would remove inter-patient variability. Some studies have shown that the concentrations of some VOCs in breath might alter following surgical resection of lung tumors and that these differences are maintained at different time points (Poli et al., 2005, 2008). However, other studies have shown that the VOCs' profile does not seem to change prior o to and after tumor resection (Dent et al., 2013; Horváth et al., 2009; Phillips, 2007). Further work is needed to understand better whether breath analysis could be useful to assess tumor response to treatment. Interestingly, increased levels of H_2O_2 concentrations from oxidative stress were seen in lobectomy patients when compared with pneumonectomy patients (Lases et al., 2000). Thus, changes in metabolism following surgery and/or other therapeutic options (chemotherapy, radiotherapy, targeted therapies) could produce volatiles, which needs to be taken into account. This certainly raises the issue that some of VOCs measured in breath following certain treatments (chemotherapy, radiotherapy, etc.) might not be directly produced by cancer cells.

3.3.3   FOLLOW-UP

Once patients have been diagnosed and treated, they need to be followed-up, and several tests be carried out regularly in order to confirm that the disease is not coming back. As described above, these tests could be invasive, costly and give radiation. Therefore, VOCs analysis could also have a role in patients follow up. The advantage here when compared to using this methodology to diagnose cancer is that a VOCs' profile of the patient with cancer is already available. Furthermore, a VOCs profile of the same patient during and after treatment will also be available. Thus, at follow-up, special attention can be placed on those volatiles that might have been raised prior to treatment and their levels diminished following treatment.

3.4   CLINICAL IMPLEMENTATION

The translation of new technologies into clinical practice has to evaluate questions such as:

  i)   What are the most achievable, strategic target applications?
 ii)   What are the technical challenges, and how can they be addressed? and
iii)   What are the challenges to implementation (legislative, clinical trials, etc.), and how can they be addressed? (Byrne et al., 2015).

While several studies have shown that the measurement of VOCs can identify cancer both *in vitro* and *in vivo*, this has not translated yet into a clear clinical application of breath analysis in the management of cancer patients. For this to happened, several issues will need to be tackled (Smith and Španěl, 2015). These include:

- **Sampling methods.** The concentrations of volatile compounds in exhaled breath may depend on the sampling method (Di Francesco et al., 2008; Herbig et al., 2008; Miekisch et al., 2008; Filipiak et al., 2012). The present methods of sampling include direct on-line sampling, sample collection into flexible or inflexible containers, and direct trapping and pre-concentration (Smith and Španěl, 2015). One of the advantages of direct sampling is to avoid the loss of trace compounds. However, the measurement of VOCs is limited to the exhalation time. Sample collection into containers such as bags allows for sample storage. However, losses of trace compounds might occur. Furthermore, storage of breath samples is limited. Direct trapping and pre-concentration from exhaled breath samples onto an absorbent might include certain uncertainty in adsorption and desorption efficiencies (Smith and Španěl, 2015). Some compounds might result from artifacts such as contamination, degradation or oxidation, which can occur during collection, storage or measurement (de Lacy Costello et al., 2014). Furthermore, breaths may considerably vary from each other due to different modes and depth of breathing, so obtaining multiple breaths may be preferable in order to acquire breath samples that are highly reproducible (Lourenço and Turner, 2014). All this indicates that there is no ideal method to sample breath yet and that further work is needed to better identify the preferred method of breath sampling for each individual situation.

- **Instrumentation.** The final aim is to use instruments for breath analysis, which will help in the management of cancer patients (and other non-malignant diseases). Such apparatus have to be acceptable to patients, especially in those special circumstances involving patients in respiratory distress or under ventilation. The apparatus must be safe and should comply with appropriate infection control requirements for use in the clinical environment (Rattray et al., 2014). From the operator point of view, it needs to be user-friendly, as automated as possible and provide a high-throughput. Also, different instruments might provide different results but these potential differences have to be clinically acceptable with or without the use of correction factors between instruments. More important, data has to be transferrable between instruments. Costs are also an important issue to be taken into account especially for the application of this technology in developing countries. Instruments will also need to be trialed and validated. Also, reproducibility of systems is a vital technical challenge to be addressed, as is transferability of datasets between systems (Byrne et al., 2015).
- **Data.** The use of VOCs as biomarkers will probably have to include system biology approaches in addition to chemometrics (Subramaniam et al., 2013). The way data is managed (correction, normalization, analysis) can vary markedly from different studies and on the instrument used. It is important to be careful and fully understand the way data has been corrected and/or analyzed, so the conclusions are sound. Studies with significant sample sizes through multicenter studies are desirable.

Prior to a full clinical implementation, breath collection and analysis procedures will need to be standardized. Furthermore, a better understanding of the nature and origin of volatile biomarkers together with a better understanding of the potential confounders will also be required before achieving a diagnosis with high sensitivity and specificity (van der Schee et al., 2015).

3.5 CONCLUDING REMARKS

The work carried out so far on breath analysis for the management of cancer has shown that the proof of concept has already been achieved. However, there are still challenges facing this field, and further work is needed to develop this technology into a clinical application in the management of cancer. Engagement between academic researchers and clinicians is already present. The expertise brought by them together with instrument developers, data analysts and others will help to pave the way to develop this field into clinics. This will contribute to tackle issues such as better understanding the origin of VOCs, and the possible differences caused by instrumentation, sample collection and storage, and data analysis. This should ultimately lead breath analysis as the so-called "breath fingerprinting" towards a better management of patients with cancer.

KEYWORDS

- cancer
- clinical implementation
- ex vivo
- in vitro
- in vivo
- management
- VOCs

REFERENCES

Adiguzel, Y., & Kulah, H., (2015). Breath sensors for lung cancer diagnosis. *Biosens. Bioelectron.*, *65*, 121–138.

Alix-Panabieres, C., & Pantel, K., (2013). Circulating tumor cells: liquid biopsy of cancer. *Clin. Chem.*, *59*, 110–118.

Altomare, D. F., Lena, M. D., Porcelli, F., Trizio, L., Travaglio, E., Tutino, M., Dragonieri, S., Memeo, V., & De Gennaro, G., (2013). Exhaled volatile organic compounds identify patients with colorectal cancer. *Br. J. Surg.*, *100*, 144–150.

Amal, H., Leja, M., Broza, Y. Y., Tisch, U., Funka, K., Liepniece-Karele, I., Skapars, R., Xu, Z. Q., Liu, H., & Haick, H., (2013). Geographical variation in the exhaled volatile organic compounds. *J. Breath Res.*, *7*, 047102.

Amal, H., Leja, M., Funka, K., Skapars, R., Sivins, A., Ancans, G., Liepniece-Karele, I., Kikuste, I., Lasina, I., & Haick, H., (2016). Detection of pre-cancerous gastric lesions and gastric cancer through exhaled breath. *Gut*, *65*, 400–407.

Amann, A., Miekisch, W., Pleil, J., Risby, T., & Schubert, J., (2010). Methodological issues of sample collection and analysis of exhaled breath. In: *Exhaled Biomarkers*, Horvath, I., De Jongste, J. C., Eds. European Respiratory Society: Plymouth, UK, pp. 96–107.

Barash, O., Zhang, W., Halpern, J. M., Hua, Q. L., Pan, Y. Y., Kayal, H., Khoury, K., Liu, H., Davies, M. P., & Haick, H., (2015). Differentiation between genetic mutations of breast cancer by breath volatolomics, *Oncotarget*, *6*, 44864–44876.

Beauchamp, J. D., & Pleil, J. D., (2013). Simply breath-taking? Developing a strategy for consistent breath sampling. *J. Breath Res.*, *7*, 042001.

Blumenthal, R. D., & Goldenberg, D. M., (2007). Methods and goals for the use of *in vitro* and *in vivo* chemosensitivity testing. *Mol. Biotechnol.*, *35*, 185–197.

Brunner, C., Szymczak, W., Höllriegl, V., Mörtl, S., Oelmez, H., Bergner, A., Huber, R. M., Hoeschen, C., & Oeh, U., (2010). Discrimination of cancerous and non-cancerous cell lines by headspace-analysis with PTR-MS. *Anal. Bioanal. Chem.*, *397*, 2315–2324.

Buszewski, B., Ulanowska, A., Kowalkowski, T., & Cieliski, K., (2012). Investigation of lung cancer biomarkers by hyphenated separation techniques and chemometrics. *Clin. Chem. Lab. Med.*, *50*, 573–581.

Byrne, H., Baranska, M., Puppels, G., Stone, N., Wood, B., Gough, K. M., Lasch, P., Heraud, P., Sulé-Suso, J., & Sockalingum, G. D., (2015). Spectrum pathology for the next generation: quo vadis? *Analyst*, *140*, 2066–2073.

Cao, W., & Duan, Y., (2006). Breath analysis: Potential for clinical diagnosis and exposure assessment. *Clin. Chem.*, *52*, 800–811.

Chang, J. C., Wooten, E. C., Tsimelzon, A., Hilsenbeck, S. G., Gutiérrez, M. C., Elledge, R., Mohsin, S., Osborne, C. K., Chamness, G. C., Allred, D. C., & O'Connell, P., (2003). Gene expression profiling for the prediction of therapeutic response to docetaxel in patients with breast cancer. *Lancet*, *362*, 362–369.

Chen, J., Wang, W., Lv, S., Yin, P., Zhao, X., Lu, X., Zhang, F., & Xu, G., (2009). Metabonomics study of liver cancer based on ultra-performance liquid chromatography coupled to mass spectrometry with HILIC and RPLC separations. *Anal. Chim. Acta*, *650*, 3–9.

Corradi, M., Poli, D., Banda, I., Bonini, S., Mozzoni, P., Pinelli, S., Alinovi, R., Andreoli, R., Ampollini, L., Casalini, A., Carbognani, P., Goldoni, M., & Mutti, A., (2015). Exhaled breath analysis in suspected cases of non-small cell lung cancer: a cross-sectional study. *J. Breath Res.*, *9*, 027101.

Cree, I. A., (2009). Chemosensitivity and chemoresistance testing in ovarian cancer. *Curr. Opin. Obstet. Gynecol.*, *21*, 39–43.

De Lacy Costello, B., Amann, A., Al-Kateb, A., Flynn, C., Filipiak, W., Khalid, T., Osborne, D., & Ratcliffe, N. M., (2014). A review of the volatiles from the healthy human body. *J. Breath Res.*, *8*, 0140001.

Deng, C., Zhang, X., & Li, N., (2004a). Investigation of volatile biomarkers in lung cancer blood using solid-phase microextraction and capillary gas chromatography-mass spectrometry. *J. Chromatogr. B*, *808*, 269–277.

Deng, C., Zhang, X., & Li, N., (2004b). Development of headspace solid-phase microextraction with on-fiber derivatization for determination of hexanal and heptanal in human blood. *J. Chromatogr. B.*, *813*, 47–52.

Dent, A. G., Sutedja, T. G., & Zimmerman, P. V., (2013). Exhaled breath analysis for lung cancer. *J. Thorac. Dis.*, *5*, 540–550.

Di Francesco, F., Fuoco, R., Trivella, M. G., & Ceccarini, A., (2005). Breath analysis: trends in techniques and clinical applications. *Microchem. J.*, *79*, 405–410.

Di Francesco, F., Loccioni, C., Fioravanti, M., et al., (2008). Implementation of Fowler's method for end-tidal air sampling. *J. Breath Res.*, *2*, 037009.

Ding, X., & Kaminsky, L. S., (2003). Human extrahepatic cytochromes P450: function in xenobiotic metabolism and tissue-selective chemical toxicity in the respiratory and gastrointestinal tracts. *Annu. Rev. Pharmacol. Toxicol.*, *43*, 149–173.

Dragonieri, S., Schot, R., Mertens, B. J., Le Cessie, S., Gauw, S. A., Spanevello, A., Resta, O., Willard, N. P., Vink, T. J., Rabe, K. F., Bel, E. H., & Sterk, P. J., (2007). An electronic nose in the discrimination of patients with asthma and controls. *J. Allergy Clin. Immunol.*, *120*, 856–862.

Filipiak, W., Filipiak, A., Ager, C., Wiesenhofer, H., & Amann, A., (2012). Optimization of sampling parameters for collection and preconcentration of alveolar air by needle traps. *J. Breath Res.*, *6*, 027107.

Filipiak, W., Filipiak, A., Sponring, A., Schmid, T., Zelger, B., Ager, C., Klodzinska, E., Denz, H., Pizzini, A., Lucciarini, P., Jamnig, H., Troppmair, J., & Amann, A., (2014). Comparative analyses of volatile organic compounds (VOCs) from patients, tumors and transformed cell lines for the validation of lung cancer-derived breath markers. *J. Breath Res.*, *8*, 027111.

Filipiak, W., Sponring, A., Filipiak, A., Ager, C., Schubert, J., Miekisch, W., Amann, A., & Troppmair, J., (2010). TD-GC-MS analysis of volatile metabolites of human lung cancer and normal cells *in vitro*. *Cancer Epidemiol. Biomark. Prev.*, *19*, 182–195.

Filipiak, W., Sponring, A., Mikoviny, T., Ager, C., Schubert, J., Miekisch, W., Amann, A., & Troppmair, J., (2008). Release of volatile organic compounds (VOCs) from the lung cancer cell line CALU-1 *in vitro*. *Cancer Cell Int.*, *8*, 17.

Gatenby, R. A., & Gillies, R. J., (2004). Why do cancers have high aerobic glycolysis? *Nature Rev.*, *4*, 891–899.

Ghezzi, P., Jaquet, V., Marcucci, F., & Schmidt, H. H., (2017). The oxidative stress theory of disease: levels of evidence and epistemological aspects. *Br. J. Pharmacol.*, *174*, 1784–1796.

Gisbert, J. P., & Pajares, J. M., (2004). ^{13}C-Urea breath test in the diagnosis of Helicobacter pylori infection - a critical review. *Aliment. Pharmacol. Ther.*, *20*, 1001–1017.

Guadagni, R., Miraglia, N., Simonelli, A., Silvestre, A., Lamberti, M., Feola, D., Acampora, A., & Sannolo, N., (2011). Solid-phase microextraction-gas chromatography-mass spectrometry method validation for the determination of endogenous substances: urinary hexanal and heptanal as lung tumor biomarkers. *Anal. Chim. Acta*, *701*, 29–36.

Haick, H., Broza, Y. Y., Mochalski, P., Ruzsanyi, V., & Amann, A., (2014). Assessment, origin, and implementation of breath volatile cancer markers. *Chem. Soc. Rev.*, *43*, 1423–1449.

Hakim, M., Broza, Y. Y., Barash, O., Peled, N., Phillips, M., Amann, A., & Haick, H., (2012). Volatile organic compounds of lung cancer and possible biochemical pathways. *Chem. Reviews*, *112*, 5949–5966.

Hamburger, A. W., & Salmon, S. E., (1977). Primary bioassay of human tumor stem cells. *Science, 197*, 461–463.

Hensing, T. A., & Salgia, R., (2013). Molecular biomarkers for future screening of lung cancer. *J. Surg. Oncol., 108*, 327–333.

Herbig, J., Titzmann, T., Beauchamp, J., Kohl, I., & Hansel, A., (2008). Buffered end-tidal (BET) sampling – a novel method for real-time breath-gas analysis. *J. Breath Res., 2*, 037008.

Horváth, I., Lázár, Z., Gyulai, N., Kollai, M., & Losonczy, G., (2009). Exhaled biomarkers in lung cancer. *Eur. Respir. J., 34*, 261–275.

Huang, J., Kumar, S., Abbassi-Ghadi, N., Španěl, P., Smith, D., & Hanna, G. B., (2013). Selected ion flow tube mass spectrometry analysis of volatile metabolites in urine headspace for the profiling of gastro-oesophageal cancer. *Anal. Chem., 85*, 3409–3416.

Huang, X., Gao, P., Sun, J., Chen, X., Song, Y., Zhao, J., Xu, H., & Wang, Z., (2015). Clinico-pathological and prognostic significance of circulating tumor cells in patients with gastric cancer: a meta-analysis. *Int. J. Cancer 136*, 21–33.

Jobu, K., Sun, C., Yoshioka, S., Yokota, J., Onogawa, M., Kawada, C., Inoue, K., Shuin, T., Sendo, T., & Miyamura, M., (2012). Metabolomics study on the biochemical profiles of odor elements in urine of human with bladder cancer. *Biol. Pharm. Bulletin, 35*, 639–642.

Jones, R. G., & Thompson, C. B., (2009). Tumor suppressors and cell metabolism: a recipe for cancer growth. *Genes. Dev., 23*, 537–548.

Jorgensen, C., (2009). Link between cancer stem cells and adult mesenchymal stromal cells: implications for cancer therapy. *Regen. Med., 4*, 149–152.

Kalluri, U., Naiker, M., & Myers, M. A., (2014). Cell culture metabolomics in the diagnosis of lung cancer - the influence of cell culture conditions. *J. Breath Res., 8*, 027109.

Katoh, S., Goi, T., Naruse, T., Ueda, Y., Kurebayashi, H., Nakazawa, T., Kimura, Y., Hirono, Y., & Yamaguchi, A., (2015). Cancer stem cell marker in circulating tumor cells: expression of CD44 variant exon 9 is strongly correlated to treatment refractoriness, recurrence and prognosis of human colorectal cancer. *Anticancer Res., 35*, 239–244.

Khalid, T. Y., Saad, S., Greenman, J., De Lacy Costello, B., Probert, C. S., & Ratcliffe, N. M., (2013). Volatiles from oral anaerobes confounding breath biomarker discovery. *J. Breath Res., 7*, 017114.

King, J., Koch, H., Unterkofler, K., Mochalski, P., Kupferthaler, A., Teschl, G., Teschl, S., Hinterhuber, H., & Amann, A., (2010). Physiological modelling of isoprene dynamics in exhaled breath. *J. Theor. Biol., 267*, 626–637.

King, J., Kupferthaler, A., Unterkofler, K., Koc, H., Teschl, S., Teschl, G., Miekisch, W., Schubert, J., Hinterhuber, H., & Amann, A., (2009). Isoprene and acetone concentration profiles during exercise in an ergometer. *J. Breath Res., 3*, 027006.

Kischkel, S., Miekisch, W., Sawacki, A., Straker, E. M., Trefz, P., Amann, A., & Schubert, J. K., (2010). Breath biomarkers for lung cancer detection and assessment of smoking related effects - confounding variables, influence of normalization and statistical algorithms. *Clin. Chim. Acta, 411*, 1637–1644.

Kouremenos, K. A., Johansson, M., & Marriott, P. J., (2012). Advances in gas chromatographic methods for the identification of biomarkers in cancer. *J. Cancer, 3*, 404–420.

Krilaviciute, A., Heiss, J. A., Leja, M., Kupcinskas, J., Haick, H., & Brenner, H., (2015). Detection of cancer through exhaled breath: a systematic review. *Oncotarget, 6*, 38643–38657.

Kumar, S., Huang, J. Z., Abbassi-Ghadi, N., Španěl, P., Smith, D., & Hanna, G. B., (2013). Selected ion flow tube mass spectrometry analysis of exhaled breath for volatile organic compound profiling of esophagogastric cancer. *Anal. Chem.*, *85*, 6121–6128.

Kwak, J., Gallagher, M., Ozdener, M. H., Wysocki, C. J., Goldsmith, B. R., Isamah, A., Faranda, A., Fakharzadeh, S. S., Herlyn, M., Johnson, A. T., & Prêti, G., (2013). Volatile biomarkers from human melanoma cells. *J. Chromatogr. B., 931*, 90–96.

Laser, E. C., Duurkens, V. A. N., Gerritsen, W. B., & Haas, F. J., (2000). Oxidative stress after lung resection therapy: a pilot study. *Chest, 117*, 999–1003.

Lechner, M., Moser, B., Niederseer, D., Karlseder, A., Holzknecht, B., Fuchs, M., Colvin, S., Tilg, H., & Rieder, J., (2006). Gender and age specific differences in exhaled isoprene levels. *J. Respir. Physiol. Neurobiol., 154*, 478–483.

Liberti, M. V., & Locasale, J. W., (2016). The Warburg effect: How does it benefit cancer cells? *Trends Biochem. Sci., 41*, 211–218.

Logan, R. P. H., (1998). Urea breath tests in the management of Helicobacter pylori infection. *Gut, 43*, 47–50.

Lourenço, C., & Turner, C., (2014). Breath analysis in disease diagnosis: Methodological considerations and applications. *Metabolites, 4*, 465–498.

Manolis, A., (1983). The diagnostic potential of breath analysis. *Clin. Chem., 29*, 5–15.

Mazzone, P. J., Hammel, J., Dweik, R., Na, J., Czich, C., Laskowski, D., & Mekhail, T., (2007). Diagnosis of lung cancer by the analysis of exhaled breath with a colorimetric sensor array. *Thorax, 62*, 565–568.

McDonald, S. A., Graham, T. A., Schier, S., Wright, N. A., & Alison, M. R., (2009). Stem cells and solid cancers. *Virchows Arch., 455*, 1–13.

Miekisch, W., Kischkel, S., Sawacki, A., Liebau, T., Mieth, M., & Schubert, J. K., (2008). Impact of sampling procedures on the results of breath analysis. *J. Breath Res., 2*, 026007.

Miekisch, W., Schubert, J. K., & Noeldge-Schomburg, G. F., (2004). Diagnostic potential of breath analysis: focus on volatile organic compounds. *Clin. Chim. Acta, 347*, 25–39.

Mochalski, P., King, J., Klieber, M., Unterkofler, K., Hinterhuber, H., Baumann, M., & Amann, A., (2013). Blood and breath levels of selected volatile organic compounds in healthy volunteers. *Analyst, 138*, 2134–2145.

Mukhopadhyay, R., (2004). Don't waste your breath. Researchers are developing breath tests for diagnosing diseases, but how well do they work? *Anal. Chem., 76*, 273–276.

Murray, R., Granner, D., Mayes, P., & Rodwell, V., (2006). *Harper's Illustrated Biochemistry, 27 edn.*, McGraw-Hill Medical, New York, ISBN-10: 0071461973, pp. 672.

Notter, R. H., (2000). Lung surfactants. Basic science and clinical applications. In: *Lung Biology in Health and Disease, 149*, CRC Press: New York, ISBN 10: 0824704010, ISBN 13: 9780824704018, pp. 464.

Ochs, M., Nyengaard, J. R., Jung, A., Knudsen, L., Voigt, M., Wahlers, T., Richter, J., & Gundersen, H. J., (2004). The number of alveoli in the human lung. *Am. J. Respir. Crit. Care Med., 169*, 120–124.

Pauling, L., Robinson, A. B., Teranishi, R., & Cary, P., (1971). Quantitative analysis of urine vapour and breath by gas-liquid partition chromatography. *Proc. Natl. Acad. Sci. USA, 68*, 2374–2376.

Peled, N., Hakim, M., Bunn, P. A., Jr, Miller, Y. E., Kennedy, T. C., Mattei, J., Mitchell, J. D., Hirsch, F. R., & Haick, H., (2012). Non-invasive breath analysis of pulmonary nodules. *J. Thorac. Oncol., 7*, 1528–1533.

Peng, G., Hakim, M., Broza, Y. Y., Billan, S., Abdah-Bortnyak, R., Kuten, A., Tisch, U., & Haick, H., (2010). Detection of lung, breast, colorectal, and prostate cancers from exhaled breath using a single array of nanosensors. *Br. J. Cancer, 103*, 542–551.

Phillips, M., Altorki, N., Austin, J. H., Cameron, R. B., Cataneo, R. N., Greenberg, J., Kloss, R., Maxfield, R. A., Munawar, M. I., Pass, H. I., Rashid, A., Rom, W. N., & Schmitt, P., (2007). Prediction of lung cancer using volatile biomarkers in breath. *Cancer Biomark., 3*, 95–109.

Phillips, M., Cataneo, R. N., Chaturvedi, A., Kaplan, P. D., Libardoni, M., Mundada, M., Patel, U., & Zhang, X., (2013). Detection of an extended human volatome with comprehensive two-dimensional gas chromatography time-of-flight mass spectrometry. *PLoS One, 8*, e75274.

Phillips, M., Cataneo, R. N., Cummin, A. R. C., Gagliardi, A. J., Gleeson, K., Greenberg, J., Maxfield, R. A., & Rom, W. N., (2003a). Detection of lung cancer with volatile markers in the breath. *Chest, 123*, 2115–2123.

Phillips, M., Cataneo, R. N., Ditkoff, B., Fisher, P., Greenberg, J., Gunawardena, R., Kwon, C. S., Rahbari-Oskoui, F., & Wong, C., (2003b). Volatile markers of breast cancer in the breath. *Breast J., 9*, 184–191.

Phillips, M., Cataneo, R. N., Greenberg, J., Gunawardena, R., Naidu, A., & Rahbari-Oskoui, F., (2000). Effect of age on the breath methylated alkane contour, a display of apparent new markers of oxidative stress. *J. Lab. Clin. Med., 136*, 243–249.

Phillips, M., Cataneo, R. N., Saunders, C., Hope, P., Schmitt, P., & Wai, J., (2010). Volatile biomarkers in the breath of women with breast cancer. *J. Breath Res., 4*, 026003.

Poli, D., Carbognani, P., Corradi, M., Goldoni, M., Acampa, O., Balbi, B., Bianchi, L., Rusca, M., & Mutti, A., (2005). Exhaled volatile organic compounds in patients with non-small cell lung cancer: cross sectional and nested short-term follow-up study. *Respir. Res., 6*, 71.

Poli, D., Goldoni, M., Caglieri, A., Ceresa, G., Acampa, O., Carbognani, P., Rusca, M., & Corradi, M., (2008). Breath analysis in non small cell lung cancer patients after surgical tumor resection. *Acta Biomed., 79*(Suppl 1), 64–72.

Rattray, N. J. W., Hamrang, Z., Trivedi, D. K., Goodacre, R., & Fowler, S. J., (2014). Taking your breath away: Metabolomics breathes life in to personalized medicine. *Trends Biotechnol., 32*, 538–48.

Robert, J., Vekris, A., Pourquier, P., & Bonnet, J., (2004). Predicting drug response based on gene expression. *Crit. Rev. Oncol. Hematol., 51*, 205–527.

Romagnuolo, J., Schiller, D., & Bailey, R. J., (2002). Using breath tests wisely in a gastroenterology practice: an evidence-based review of indications and pitfalls in interpretation. *Am. J. Gastroenterol., 97*, 1113–1126.

Rosias, P. P. R., Dompeling, E., Dentener, M. A., Pennings, H. J., Hendriks, H. J., Van Iersel, M. P., & Jöbsis, Q., (2004). Childhood asthma: exhaled markers of airway inflammation, asthma control score, and lung function tests. *Pediatr. Pulm., 38*, 107–14.

Rossi, A., Maione, P., Colantuoni, G., Gaizo, F, D., Guerriero, C., Nicolella, D., Ferrara, C., & Gridelli, C., (2005). Screening for lung cancer: New horizons? *Crit. Rev. Oncol. Hemat., 56*, 311–320.

Rutter, A. V., Chippendale, T., Yang, Y., Španěl, P., Smith, D., & Sulé-Suso, J., (2013). Quantification by SIFT-MS of acetaldehyde released by lung cells in a 3D model, *Analyst., 138*, 91–95.

Rutter, A. V., Siddique, M. R., Filik, J., Sandt, C., Dumas, P., Cinque, G., Sockalingum, G. D., Yang, Y., & Sulé-Suso, J., (2014). Study of gemcitabine-sensitive/resistant cancer cells by cell cloning and synchrotron FTIR microspectroscopy cytometry part A, *85A*, 688–697.

Saad, F., & Pantel, K., (2012). The current role of circulating tumor cells in the dioagnosis and management of bone metastases in advanced prostate cancer. *Future Oncol.*, *8*, 321–331.

Sapoval, B., Filoche, M., & Weibel, E. R., (2002). Smaller is better—but not too small: a physical scale for the design of the mammalian pulmonary acinus. *Proc. Natl. Acad. Sci. USA*, *99*, 10411–10416.

Schmidt, K., & Podmore, I., (2015). Current challenges in volatile organic compounds analysis as potential biomarkers of cancer, *J. Biomark.*, 981458.

Schubert, J. K., Miekisch, W., Geiger, K., & Noldge-Schomburg, G. F., (2004). Breath analysis in critically ill patients: potential and limitations. *Expert. Rev. Mol. Diagn.*, *4*, 619–629.

Silva, C. L., Passos, M., & Câmara, J. S., (2012). Solid phase microextraction, mass spectrometry and metabolomic approaches for detection of potential urinary cancer biomarkers—a powerful strategy for breast cancer diagnosis. *Talanta*, *89*, 360–368.

Smith, D., & Španěl, P., (2005). Selected ion flow tube mass spectrometry (SIFT-MS) for on-line trace gas analysis. *Mass Spectrom. Rev.*, *24*, 661–700.

Smith, D., & Španěl, P., (2015). Pitfalls in the analysis of volatile breath biomarkers: suggested solutions and SIFT-MS quantification of single metabolites. *J. Breath Res.*, *9*, 022001.

Smith, D., Chippendale, T. W. E., Dryahina, K., & Španěl, P., (2013). SIFT-MS analysis of nose-exhaled breath, mouth contamination and the influence of exercise. *Curr. Anal. Chem.*, *9*, 565–575.

Smith, D., Španěl, P., Enderby, B., Lenney, W., Turner, C., & Davies, S. J., (2010). Isoprene levels in the exhaled breath of 200 healthy pupils within the age range 7–18 years studied using SIFT-MS. *J. Breath Res.*, *4*, 1–7.

Španěl, P., Turner, C., Wang, T., Bloor, R., & Smith, D., (2006). Generation of volatile compounds on mouth exposure to urea and sucrose: Implications for exhaled breath analysis. *Physiol. Meas.*, *27*, 7–17.

Sponring, A., Filipiak, W., Ager, C., Schubert, J., Miekisch, W., Amann, A., & Troppmair, J., (2010). Analysis of volatile organic compounds (VOCs) in the headspace of NCI-H1666 lung cancer cells. *Cancer Biomark.*, *7*(3), 153–161.

Sponring, A., Filipiak, W., Mikoviny, T., Ager, C., Schubert, J., Miekisch, W., Amann, A., & Troppmair, J., (2009). Release of volatile organic compounds from the lung cancer cell line NCI-H2087 *in vitro*. *Antican. Res.*, *29*, 419–426.

Stick, S. M., (2002). Non-invasive monitoring of airway inflammation. *Med. J. Aust.*, *177*, 59–60.

Stockley, R. A., (2007). Biomarkers in COPD: time for a deep breath. *Thorax*, *62*, 657–660.

Subramaniam, S., Thakur, R. K., Yadav, V. K., Nanda, R., Chowdhury, S., & Agrawal, A., (2013). Lung cancer biomarkers: state of the art. *J. Carcinog.*, *12*, 3.

Sun, X., Shao, K., & Wang, T., (2016). Detection of volatile organic compounds (VOCs) from exhaled breath as non-invasive methods for cancer diagnosis. *Anal. Bioanal. Chem.*, *408*, 2759–2780.

Tang, L., Zhao, S., Liu, W., Parchim, N. F., Huang, J., Tang, Y., Gan, P., & Zhong, M., (2013). Diagnostic accuracy of circulating tumor cells detection in gastric cancer: systematic review and meta-analysis. *BMC Cancer*, *13*, 314.

Tinhofer, I., Konschack, R., Stromberger, C., Raguse, J. D., Dreyer, J. H., Jöhrens, K., Keilholz, U., & Budach, V., (2014). Detection of circulating tumor cells for prediction of recurrence after adjuvant chemoradiation in locally advanced squamous cell carcinoma of the head and neck. *Ann. Oncol.*, *25*, 2042–2047.

Toyokuni, S., (2008). Molecular mechanisms of oxidative stress induced carcinogenesis: from epidemiology to oxygenomics. *IUBMB Life*, *60*, 441–447.

Ulanowska, A., Kowalkowski, T., Hrynkiewicz, K., Jackowski, M., & Buszewski, B., (2011). Determination of volatile organic compounds in human breath for Helicobacter pylori detection by SPME-GC-MS. *Biomed. Chromatogr.*, *25*, 391–397.

Unesono, Y., Arigami, T., Kozono, T., Yanagita, S., Hagihara, T., Haraguchi, N., Matsushita, D., Hirata, M., Arima, H., Funasako, Y., Kijima, Y., Nakajo, A., Okumura, H., Ishigami, S., Hokita, S., Ueno, S., & Natsugoe, S., (2013). Clinical significance of circulating tumor cells on peripheral blood from patients with gastric cancer. *Cancer*, *119*, 3984–3991.

Van der Schee, M. P., Paff, T., Brinkman, P., Van Aalderen, W. M., Haarman, E, G., & Sterk, P. J., (2015). Breathomics in lung disease. *Chest*, *147*, 224–231.

Wang, C., Li, C., Wang, X., Shi, C., Guo, L., Luo, S., Guo, Z., Xu, G., Zhang, F., & Li, E., (2014). Non-invasive detection for colo-rectal cancer by analysis of exhaled breath. *Anal. Bioanal. Chem.*, *406*, 4757–4763.

Wang, S., Zheng, G., Cheng, B., Chen, F., Wang, Z., Chen, Y., Wang, Y., & Xiong, B., (2014). Circulating tumor cells (CTCs) detected by PCR-RT and its prognostic role in gastric cancer: a meta-analysis of published literature. *PLoS One*, *9*, e99259.

Warburg, O., (1956). On the origin of cancer cells. *Science*, *123*, 309–314.

Wehinger, A., (2007). Lung cancer detection by proton transfer reaction mass spectrometric analysis of human breath gas. *Int. J. Mass Spectrom.*, *265*, 49–59.

WHO, (2016), Cancer factsheet. http://www.who.int/mediacentre/factsheets/fs297/en/ (accessed August 12, 2016).

Xue, R., Dong, L., Zhang, S., Deng, C., Liu, T., Wang, J., & Shen, X., (2008). Investigation of volatile biomarkers in liver cancer blood using solid-phase microextraction and gas chromatography/mass spectrometry. *Rapid Comm. Mass Spectrom.*, *22*, 1181–1186.

Zhang, Z. Y., & Ge, H. Y., (2013). Micrometastases in gastric cancer. *Cancer Lett.*, *336*, 34–45.

Zhang, Z., Ramnath, N., & Nagrath, S., (2015). Current status of CTCs as liquid biopsy in lung cancer and future directions. *Front. Oncol.*, *5*, 209.

Zimmermann, D., Hartmann, M., Moyer, M. P., Nolte, J., & Baumbach, J. I., (2007). Determination of volatile products of human colon cell line metabolism by GC-MS analysis. *Metabolomics*, *3*, 13–17.

CHAPTER 4

ARTIFICIAL OLFACTORY SYSTEMS CAN DETECT UNIQUE ODORANT SIGNATURE OF CANCEROUS VOLATILE COMPOUNDS

RADU IONESCU

*Universitat Rovira i Virgili, Department of Electronic Engineering,
Av. Paīsos Catalans 26, 43007 Tarragona, Spain,
Tel.: +34 977 55 87 54, Fax: +34 999 55 96 05,
E-mail: radu.ionescu@urv.cat*

ABSTRACT

This chapter briefly presents the concept of artificial olfactory systems and reviews the research performed so far in the field of cancer diagnosis with artificial olfactory systems by measuring the odorant signature in different body fluids (breath, urine, feces, skin, and cancerous cells).

4.1 INTRODUCTION

Artificial olfactory systems (also broadly referred as "electronic nose systems" or "e-noses") are complementary analysis tools to the analytical techniques described in the previous chapters that can be employed for the analysis of complex gaseous mixtures. The main distinction with the analytical instruments is that the artificial olfactory systems cannot identify the specific components of a gaseous mixture, but the overall signature of the odor. However, they present some paramount properties, such as they are responsive to compounds at concentration levels below the limit of detection or limit of quantification of the analytical tools, they are much simpler, easier to operate, smaller and, not of less importance, much less expensive. Furthermore, they can be made portable and require only minimal training for operation.

An artificial olfactory system is composed of an array of chemical gas sensors that is trained through robust pattern recognition algorithms to recognize specific odors to which it is exposed, after it has been previously trained to detect these odors. Each sensor in the array has different chemical characteristics with regard to the others. The overall responses of all sensors are analyzed by dedicated pattern recognition algorithms, which can be selected among a broad range of algorithms that have been specifically developed for this purpose, ranging from linear to non-linear ones, and from supervised to non-supervised [e.g., principal component analysis, discriminant function analysis, partial least squares, support vector machines, hierarchical clustering, artificial neural networks, etc. (Bishop, 2006)].

The artificial olfactory systems were developed to recognize smells that humans cannot identify. Their concept aims to mimic the human olfactory system: the sensors act in a similar way with the olfactory receptors of the human nose, while the pattern recognition algorithm plays the role of the brain. When exposed to the external odor, the sensors provide electrical signals that are analyzed and processed by the pattern recognition algorithm, which provides then a conclusion.

The sensors can base their working principle on the measurement of different physical parameters, such as electrical conductivity (chemiresistors; conducting polymers), piezoelectric properties (quartz crystal microbalance (QCM); surface acoustic wave (SAW)), optical properties (optical sensors; colorimetric sensors), or work function (field effect transistors (FET)) (Arshak et al., 2004). Figure 4.1 shows the signals acquired by a QCM sensor exposed to the breath samples of different individuals, which captured the resonance frequency shift caused by the interaction of the sensing material with the volatile organic compounds (VOCs) from the breath.

The first attempt was to develop artificial olfactory systems date from the early 1980s (Gardner and Bartlett, 1994). These system found applications in different fields, such as air quality monitoring, food and beverage industry, homeland security, and more recently in clinical research for diseases diagnosis, prognosis, and monitoring (Rock et al., 2008). In cancer diagnosis, they can be trained to detect the odorant signature of the metabolomic changes produced in the organism by the onset of the disease, by measuring the VOCs emitted by the body fluids. Although these systems are not expected to replace the current diagnostic technologies, characteristics such as low-cost (~ €10 per measurement (Rocco et al., 2015), non-invasiveness, easy operation, and portability, make them very attractive as a complementary approach to the standard methods, especially for mass-screening of high-risk population.

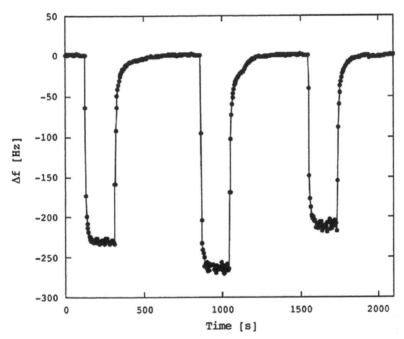

FIGURE 4.1 Signals recorded during an experiment where a QCM sensor based on Ru-TPP was exposed to three successive breath samples related to post-surgery, cancer affected, and reference breaths, respectively (From Di Natale, et al. (2003). Biosens. Bioelectron., 18(10), 1209–1218. With permission. © 2003 Elsevier.).

In view of these considerations, this chapter reviews how the application of artificial olfaction systems for diagnosis of cancer evolved in the recent years.

4.2 ARTIFICIAL OLFACTORY SYSTEMS FOR DIAGNOSIS OF CANCER

When cancer develops in the body, both the affected organ or tissue and the invading cells produce changes in the organism since the onset of the disease, causing the emission of VOCs that are different than in the normal (healthy) state by that their concentration is altered and/or new VOCs are released. As it was referred in the previous chapters, VOCs are emitted by a multitude of body fluids: cells, breath, sputum, saliva, skin (sweat), blood, urine, feces, vaginal secretion and ear wax (Broza et al., 2015; Shirasu and Touhara, 2011). This chapter will review the research performed so far in the field of cancer diagnosis with artificial olfactory systems.

4.2.1   EXHALED BREATH ANALYSIS

In breath analysis, the most commonly used artificial olfactory systems were the gold nanoparticles sensors-based e-nose NA-NOSE developed by the research group of Prof. Haick from Technion – Israel Institute of Technology, Israel (9 studies), the commercially available Cyranose 320 from Sensigent, Pasadena, US, made of conductive sensors based on carbon black mixed with different organic polymers (6 studies), and Libra-Nose, an array of QCM sensors coated with different metalloporphyrins developed by the group of Prof. Di Natale from University of Rome Tor Vergata (4 studies).

4.2.1.1   EXHALED BREATH ANALYSIS AND LUNG CANCER

The most obvious application of breath analysis is associated with the diagnosis of lung cancer and other lung diseases, as the lungs are directly connected with the airways through where the breath is exhaled. Approximately two-thirds of the breath analysis studies were conducted on lung cancer (Krilaviciute et al., 2015). The first study dates from some more than 10 years ago, when Di Natale et al. (2003) (Di Natale et al., 2003) used the LibraNose to discriminate between lung cancer patients (35 individuals) and a reference group (16 individuals); they obtained 100% correct classification of the lung cancer patients and 94% correct classification of the reference group employing the Partial Least Squares-Discriminant Analysis algorithm.

Recent reviews grouping together the research carried out since this first study in 2003 until 2015 in lung cancer diagnosis through exhaled breath analysis (Adiguzel and Kulah, 2015; Di Natale et al., 2014; Krilaviciute et al., 2015), revealed a quite limited number of studies performed in this field: 11 research papers on lung cancer diagnosis, 1 research paper on Malignant Pleural Mesothelioma diagnosis, 1 research paper on pulmonary nodules, and 1 study investigating lung cancer monitoring after lung tumor resection. The e-nose systems used in these studies were based on chemiresistive sensors, conductive polymers, QCM, SAW and colorimetric sensors (Krilaviciute et al., 2015). Part of the studies also analyzed the effect of co-founding factors (smoking habits, gender, age) or co-morbidities such as chronic obstructive pulmonary disease (COPD), head-and-neck cancer or benign medical conditions, and demonstrated that the capacity of the e-nose systems to identify lung cancer was not affected by external conditions. The accuracy, sensitivity, and specificity of the reported breath tests for lung cancer diagnosis ranged from moderate to high values (over 70% up to 100%) (Adiguzel and Kulah, 2015), although rigorous validation was not performed in all studies, which may have led to overoptimistic results (Krilaviciute et al., 2015).

The very promising results obtained in these initial studies on lung cancer diagnosis with artificial olfactory systems increased the interest of the scientific community in this field, and the diversification of the population range and scope of the analysis performed was further spread. Thus, five more studies targeting lung cancer diagnosis were published in the last year, which focused on different aspects that will be detailed in the following lines:

- Rocco et al. (2015) revealed the usefulness of the breath test for the screening of the population at high risk of developing lung cancer (Rocco et al., 2015). This study was performed on a population of 100 individuals selected based on their inclusion criteria in the high-risk category, based on age, smoking status, and chronic obstructive pulmonary disease status. The e-nose system employed was BIONOTE (biosensor-based multisensorial system for mimicking Nose, Tongue, and Eyes), developed at the University Campus Bio-Medico from Rome, Italy. BIONOTE differs from the other e-nose techniques in that it measures a set of separate transduction features (QCM and optical properties) of a class of flavonoids called anthocyanins (Santonico et al., 2013). Volunteers breath was collected with a patented device named PNEUMOPIPE (EU patent EP2641537), able to acquire the exhaled breath of an individual normally breathing into it and to collect it onto an adsorbing cartridge. The overall sensitivity and specificity of the pattern recognition model developed with the Partial Least Squares algorithm and validated using the Leave-One-Out cross validation technique, were 86% and 95%, respectively, which showed a significant difference between patients and healthy individuals. This preliminary results indicated that the BIONOTE technology might be used to reduce false-positive rates and the costs related to lung cancer screening when the low-dose computed tomography is employed – the cost of the breath test proposed in this study is around 10 €. Based on these very promising results, the authors expressed their interest in testing their model in the near future on a larger number of patients to confirm the reliability of their results.

- Gasparri et al. (2016) targeted the early diagnosis of lung cancer in its early phase (stage I) when it is easier to treat (Gasparri et al., 2016). This study was performed on 146 individuals: 70 with lung cancer confirmed by computerized tomography or positron emission tomography imaging techniques and histology (biopsy) or with clinical suspicion of lung cancer, and 76 healthy controls. The exhaled breath was analyzed with the LibraNose system, and discriminative models were built employing multivariate analysis. The sensitivity and specificity in diagnosing lung cancer were 81% and 91%, respectively. Once lung cancer was diagnosed, it was possible to discriminate Stage I versus Stages II/III/IV with 92% sensitivity and 58% specificity, respectively. The sensitivity for Stage I identification was similar for patients with or without metabolic comorbidities (90% and 94%, respectively). Actually, the e-nose provided the highest sensitivity for the identification of the subgroup of patients with Stage I lung cancer with regard to the other stages, which demonstrated its presumable suitability for the identification of lung cancer in the early phase.

- Van Hooren et al. (2016) used the AEONOSE system (formally DiagNose) based on an array of metal-oxide gas sensors, produced by the eNose company, Zutphen, the Netherlands (derived from the parent C-it BV company that developed DiagNose), to discriminate between patients with lung cancer and patients with head-and-neck squamous cell carcinoma from exhaled breath analysis (Hooren et al., 2016). The study included 78 volunteers: 34 patients with lung cancer and 53 patients with head-and-neck squamous cell carcinoma (tumor sites: oral cavity, oropharynx, nasopharynx/nasal cavity, hypopharynx, and larynx). The discrimination model was built employing an artificial neural network and validated through leave-one-out cross-validation method, obtaining a discrimination accuracy between lung cancer and head-and-neck squamous cell carcinoma of 85%.

- Nardi-Agmon et al. (2016) used the NA-NOSE sensors array to correlate the response of patients with advanced lung cancer to therapy (chemotherapy or targeted therapy) with the Response Evaluation Criteria in Solid Tumors (RECIST), which serves as the accepted standard to monitor the treatment efficacy in lung cancer (Nardi-Agmon et al., 2016). The response to therapy was classified in: complete response, partial response, stable disease, or progressive disease. The study was performed with 143 breath samples of 39 patients subjected to therapy. Two Discriminant Function Analysis models were built, and their classification accuracies were tested through leave-one-out cross-validation. The first model discriminated with 89% accuracy, 93% sensitivity and 85% specificity between controlled disease samples (partial response and stable disease) and baseline breath samples taken before therapy, while the second model discriminated with 92% accuracy, 100% specificity, but only 28% sensitivity, between controlled disease and progressive disease. The conclusion of this study was that breath analysis with NA-NOSE might provide to the oncologists a quicker indication of the lack of response to the anticancer treatment than the standard RECIST analysis, with the added benefit that the early recognition of treatment failure could improve patients' care.

- Finally, De Vries et al. (2015) investigated the integration of the e-nose technology with spirometry as a new approach for exhaled breath analysis (De Vries et al., 2015). This study was performed on a study group of 31 patients with lung cancer, 31 patients with COPD, 37 patients with asthma, and 41 healthy controls. The breath samples were analyzed with the SpiroNose system (an e-nose developed at the Academic Medical Centre, Amsterdam, The Netherlands, which comprises an array of 5 metal oxide semiconductor gas sensors arrays – three arrays monitoring exhaled breath and two reference arrays monitoring ambient air) placed at the rear end of the pneumotachograph. No correlation was found between sensors signals and external parameters such as exhaled volume, humidity and temperature. Importantly, exhaled data, after environmental background correction based on alveolar gradients, were highly reproducible for each sensor array. The classification of the four study groups with cross-validation employing Principal Component Analysis followed by Discriminant Analysis achieved values between 78–88%. The standardization of the integration of the e-nose technology with existing diagnostic tests, such as routine spirometry, could bring this technique more closely to the real 'point-of-care' set-up.

Other study aimed for instance the diagnosis of malignant pleural mesothelioma, an aggressive cancer that develops in the thin layer of tissue surrounding the lungs known as the pleura because of asbestos exposure, which was identified from healthy subjects with and without asbestos exposure with accuracies between 80% and 85% employing the Cyranose 320 system in a study performed on a small population comprising a group of 13 volunteers in each category (Dragonieri et al., 2012).

Furthermore, the NA-NOSE could distinguish significantly between benign and malignant pulmonary nodules, between adeno- and squamous-cell carcinomas and between early stage and advanced stage of lung cancer (88% accuracy in all cases) (Peled et al., 2012). The NA-NOSE sensor array could also distinguish between pre-surgery and post-surgery lung cancer states, as well as between pre-surgery lung cancer and benign states, while it could neither differentiate between pre-surgery and post-surgery benign states, nor between lung cancer and benign states after surgery, which indicated that the observed pattern was associated with the presence of malignant lung tumors (Broza et al., 2013). The excellent results of this proof-of-concept study realized in 2012, qualified of Potential Clinical Relevance by the Nanomedicine: Nanotechnology, Biology and Medicine, the journal that published it, initiated a large-scale clinical study for post-surgery follow-up of lung cancer patients that is currently underway.

4.2.1.2   DIAGNOSIS OF CANCERS CONNECTED WITH THE AIRWAYS

During the exhalation process, the breath comes in direct contact with the airways (nose, mouth, throat, larynx and pharynx), the lips and the salivary glands, which are the organs and tissues where head-and-neck cancer develops. Therefore, breath analysis could contain a characteristic odorant signature for this specific type of cancer.

This was firstly investigated by Hakim et al. (2011), who analyzed the breath of 62 volunteers (16 patients with head-and-neck cancer with different organs affected: larynx, oral cavity, supraglottis, maxilla, oropharynx, hypopharynx and nasopharynx, 20 patients with lung cancer, and 26 healthy controls), employing the NA-NOSE system (Hakim et al., 2011). The diagnostic accuracy was estimated using Support Vector Machines. Patients with head-and-neck cancer were distinguished from healthy controls with 95% accuracy (100% sensitivity and 92% specificity), and with 100% accuracy from lung cancer patients. However, this study was performed on a small population and needs further validation.

Gruber et al. (2014) targeted the diagnosis of head-and-neck squamous cell carcinoma (that develops in the mucous membranes of the mouth, nose, and throat) from exhaled breath analysis (Gruber et al., 2014). This study performed on a population of 62 volunteers: 22 patients with head-and-neck squamous cell carcinoma (affected organs: larynx and pharynx), 21 patients with benign tumors (the same affected organs: larynx and pharynx), and 19 healthy controls. The breath samples were analyzed with NA-NOSE. The models developed employing the Discriminant Function Analysis algorithm gave values of 77% sensitivity, 90% specificity and 83% overall classification accuracy between head-and-neck squamous cell carcinoma and patients with benign conditions. Exactly the same classification results were obtained between head-and-neck squamous cell carcinoma and healthy controls. The classification accuracy between head-and-neck squamous cell carcinoma that affected either the larynx or the pharynx was over 90%, while the classification between early-stage and late-stage head-and-neck squamous cell carcinoma reached a classification accuracy of 95%. The main conclusions of this study were that the breath test could potentially provide valuable complementary information for distinguishing malignant and benign lesions with a similar appearance, and could indicate malignancy prior to the endoscopic examination.

Almost concomitantly, Leuni et al. (2014) employed the DiagNose (array of metal oxide gas sensors developed by C-it BV, Zutphen, the Netherlands) for the diagnosis of head-and-neck squamous cell carcinoma (Leunis et al., 2014). This study was performed with 36 patients with head-and-neck squamous cell carcinoma localized in different organs (oropharynx, hypopharynx, and supraglottic larynx) and a group of 23 control patients attended for other benign conditions (ear operation, nose operation, benign neck operation, benign larynx operation, snoring operation, and tonsillectomy). Logistic regression showed a significant difference in sensors resistance patterns between patients diagnosed with head-and-neck squamous cell carcinoma and the control group, achieving a sensitivity of 90% and a specificity of 80% in head-and-neck squamous cell carcinoma identification.

Very recently, as we mentioned before, van Hooren et al. (2016) discriminated with 85% accuracy between head-and-neck squamous cell carcinoma and lung cancer by measuring patients breath samples with the AEONOSE system, in a study that involved 78 volunteers: 53 patients with head-and-neck squamous cell carcinoma (tumor sites: oral cavity, oropharynx, nasopharynx/nasal cavity, hypopharynx and larynx) and 34 patients with lung cancer (Hooren et al., 2016).

4.2.1.3   OTHER TYPES OF CANCER DIAGNOSED THROUGH EXHALED BREATH ANALYSIS

Besides the respiratory system and the airways, breath composition is likely to contain compounds that come from the digestive system. The possibility of diagnosing gastric cancer with an electronic nose system was studied only recently. Employing the NA-NOSE, Xu et al. (2013) distinguished gastric cancer from other benign gastric conditions with 89% sensitivity and 90% specificity, and early-stage gastric cancer versus late-stage gastric cancer with 89% sensitivity and 94% accuracy (models developed with discriminant function analysis and tested through leave-one-out cross-validation) (Xu et al., 2013). This study was performed on a cohort of 130 patients with gastric complaints (37 with gastric cancer, 32 with ulcer and 61 with less severe conditions). In a posterior larger scale study including 484 patients with gastric cancer and gastric intestinal metaplasia published by Amal et al. (2015), the same research group attempted the detection of precancerous gastric lesions and gastric cancer through exhaled breath analysis with the NA-NOSE (Amal et al., 2016). The samples were randomly divided into a training set (70% of samples) and validation set (30% of samples), and discriminative models were build using discriminant function analysis. The classification accuracy between gastric cancer and gastric intestinal metaplasia was 92% (73%

sensitivity, 98% specificity). The further analysis aimed the assessment of gastric intestinal metaplasia, classified accordingly to the operative link on gastric intestinal metaplasia (OLGIM) staging system, which was used to stratify the presence/absence of risk level of precancerous lesions, where patients with OLGIM stages III–IV were considered to be at high risk. The accuracy of discrimination between gastric cancer and OLGIM 0–II was 87% (97% sensitivity, 84% specificity), and between gastric cancer and OLGIM III–IV was 90% (93% sensitivity, 80% specificity), which suggested that breath analysis with NA-NOSE could provide the missing non-invasive screening tool for gastric cancer and related precancerous lesions.

On the other hand, when cancer develops into the body, it produces oxidative stress at the cellular level, which causes the emission of specific volatile biomarkers into the blood. From there, through exchange via the lungs, these biomarkers arrive at the exhaled air and are released through the breath. Therefore, exhaled breath analysis could contain odorant patterns specific to cancers developed at the internal organs, which could be detected by artificial olfactory systems. This was firstly investigated by Peng et al. (2010), who could diagnose and discriminate four different types of cancers: lung, breast, colorectal and prostate (Peng et al., 2010). The study was performed on a population of 96 volunteers: 30 primary lung cancer patients, 26 primary colon cancer patients, 22 primary breast cancer patients, 18 primary prostate cancer patients, and 22 healthy controls. The breath samples were analyzed with the NA-NOSE system, and the principal component analysis performed showed a perfect discrimination between all these categories, with just a slight overlap between the prostate cancer patients and the healthy subjects (see Figure 4.2).

In another study, Shuster et al. (2011) used the NA-NOSE to discriminate between females with breast cancer (13 subjects), females with benign breast conditions (16 subjects) and healthy female volunteers (7 subjects) (Shuster et al., 2011). Using the first component (PC1) of Principal Component Analysis as a classification tool, there were two subjects misclassified from each group between the benign and malignant breast conditions, while the group of healthy controls totally matched the group of patients with benign conditions.

More recently, Amal et al. (2015) demonstrated the possibility to discriminate females with ovarian cancer (48 subjects) from females that have no tumor (48 subjects) and from females that have benign genital tract neoplasia (86 subjects), analyzing exhaled breath samples with the NA-NOSE system (Amal et al., 2015). The discriminant function analysis models developed showed a discrimination accuracy of 89% between ovarian cancer and healthy individuals (79% sensitivity, 100% specificity), and of 71% between ovarian cancer and all the other volunteers together (71% sensitivity, 71% accuracy). The models were built with 70% of the data (randomly selected) and validated with the remaining 30% of the data. As disease diagnosis at the early stage is of highest interest to the success of the treatment, further models were developed to discriminate between early-stage and late-stage ovarian cancer (accuracy obtained: 72%), and between early-stage ovarian cancer and tumor-free volunteers (accuracy obtained: 75%).

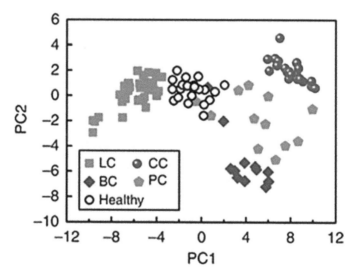

FIGURE 4.2   (See color insert.) PCA scores plot showing the discrimination between lung cancer (LC), colorectal cancer (CC), breast cancer (BC), prostate cancer (PC), and healthy controls (adapted from Peng, et al. (2010). Br. J. Cancer 103(4), 542–551. With permission.) © 2010 Nature Publishing Group.

The breath analysis studies performed so far suggested that each type of cancer (or disease in general) has its own breathprint, therefore the presence of the target disease would not be masked by the presence of other diseases if artificial olfactory systems with specific sensors and specific fingerprint matching tools are employed (Tisch et al., 2013).

4.2.2   URINE VOLATILES ANALYSIS

Urinary tract cancers (which affect the bladder, the kidney, and the tubes connected to them) are most likely to provide compounds that are released through the urine.

Urine headspace was firstly analyzed by Bernabei et al. (2008) with LibraNose, for the diagnosis of bladder and prostate cancer (Bernabei et al., 2008). This study was performed on a cohort of 117 patients affected by various urological pathologies: 25 affected by bladder cancer, 12 by prostate cancer, 29 by benign prostatic hypertrophy, 33 by other urological pathologies, and 18 healthy controls; 15 patients were measured twice, before and after surgery for cancer treatment. All measurements were performed on site in the hospital with first-morning urine samples. The samples were stored in sealed vials at 25°C before analysis. 10 mL of the headspace of each sample was extracted and injected into a 2 L sterile bag pre-filled with N2 by means of a chromatographic syringe, and the content of the bag was flown at the constant speed of 0.3 L/min into the sensors chamber. The e-nose data were processed by Principal Component Analysis (PCA) and Discriminant Analysis solved by Partial Least Squares (PLS-DA). The PLS-DA model showed a complete discrimination between healthy and ill individuals, except for the post-surgery samples, and a gradual differentiation between prostate and bladder cancer. The application of PCA to the whole dataset showed a migration of the post-surgery class towards the healthy group. These results confirmed a strict correlation between urine headspace and urological pathologies.

Early detection of prostate cancer was investigated by Asimakopoulos et al. (2014), from the same research group (Asimakopoulos et al., 2014). In this study were included 41 volunteers suspected of prostate cancer, and urine samples were collected before the biopsy. Both the initial part of the voided urine and urine midstream were collected in separate vials, and their headspace was analyzed with LibraNose immediately after collection. E-nose results were compared with the prostate biopsy pathological results. The e-nose was able to diagnose positive prostate biopsy results from negative prostate biopsy results with 71% sensitivity and 93% specificity when the first part of urine was analyzed, whereas the midstream urine couldn't correlate the samples correctly with the biopsy outcome. This study demonstrated that, if the correct part of urine is analyzed, it is possible to diagnose prostate cancer at an early stage by measuring urine headspace with an artificial olfactory system.

Roine et al. (2014) evaluated the ability of an e-nose system to discriminate prostate cancer from benign prostatic hyperplasia using urine headspace (Roine et al., 2014). The study was performed in 50 patients with prostate cancer and 15 patients with benign prostatic hyperplasia scheduled for robotic assisted laparoscopic radical prostatectomy or transurethral resection of the prostate; from the latter group, 9 patients provided urine samples also at 3 months after surgery, which served as control. Urine headspace was analyzed with the handheld ChemPro® 100-eNose (Environics Oy, Finland), consisting of an Ion Mobility Spectrometry cell and six semiconductor gas sensors. Data analysis was conducted using linear discriminant analysis and logistic regression. Using leave-one-out cross-validation, the e-nose reached a sensitivity of 78% and a specificity of 67%, which are comparable with the results of the prostate specific antigen testing, the common approach to detect prostate cancer.

Colorectal cancer typically originates in the secretory cells lining the gut. The interaction of colonic cells, human gut microflora, and invading pathogens produce a variety of volatiles within the lower gastrointestinal tract, and the products of this process could be found in urine (Arasaradnam et al., 2012). Following this idea, Westenbrink et al. (2015) investigated the possibility to detect colorectal cancer from urine headspace analysis with an e-nose device (Westenbrink et al., 2015). The study was performed on a cohort of 92 urine samples collected from pre-diagnosed patients with colorectal cancer (39 subjects), with irritable bowel syndrome (35 subjects), and healthy controls (18 subjects). The e-nose system employed was the WOLF (Warwick olfaction system) developed at University of Warwick, UK, consisting of an array of 13 electrochemical and optical sensors. Urine samples were collected and stored at −80°C within 2 h after collection, and defrosted overnight at 5°C before the experiments. 5 mL of headspace formed by heating the samples at 40°C for 5 min were introduced to the sensors system; this process was repeated three times per sample to ensure that the full profile of volatiles content of each sample was analyzed. The Linear Discriminant Analysis model developed obtained a full 3-classes classification and showed 78% sensitivity and 79% specificity at distinguishing colorectal cancer from irritable bowel syndrome by assessing the success of re-classification using the $(n-1)$ K-nearest neighbor algorithm.

4.2.3   FECES ANALYSIS

The waste products of the digested food and liquids pass through the colon and rectum before being evacuated from the body as feces, therefore these can contain compounds produced by the intestinal microbiota in the colon as well as by metabolic processes at the cellular level, and are an important source for detecting specific traits associated with colorectal cancer (Narayanan et al., 2014). The analysis of feces headspace with an e-nose system for the diagnosis of colorectal cancer was studied by De Meij et al. (2014) (Meij et al., 2014). Fecal samples were collected from patients scheduled to undergo elective colonoscopy; the exclusion criteria comprised patients with a history of inflammatory bowel disease and patients with inadequate bowel cleansing. The study group comprised 40 patients with colorectal cancer, 60 patients with advanced adenomas (together 100 patients with advanced neoplasia), and 57 controls. The patients collected themselves at home their fecal samples and stored them in their fridge before taking them to the hospital, where they were stored at −20°C before analysis. Approximately 2 g of frozen feces was transferred into a sealed vacutainer, where they were gradually heated to 37°C for 1 h to enhance VOCs release from the stools. The headspace was measured with the Cyranose 320 system for 1 min, and classification models were built employing the Canonical Discriminant Analysis. 85% sensitivity and 87% specificity were achieved at the discrimination between colorectal carcinoma vs. control; 62% sensitivity and 86% specificity between advanced adenoma vs. control; 75% sensitivity and 73% specificity between advanced adenomas vs. colorectal carcinoma; and 85% sensitivity and 68% specificity between advanced neoplasia vs. control. The intraclass correlation coefficients for five duplicate measurements of three distinct fecal samples were 0.997 in all cases, indicating extremely high levels of measurement repeatability. The results of this proof of concept study suggested that fecal headspace analysis with artificial olfactory systems seems to hold a promising potential for early detection of colorectal cancer and advanced neoplasia.

4.2.4   SKIN HEADSPACE ANALYSIS

The skin releases volatiles produced mainly by the skin glands and the bacterial populations at the skin surface, but also by other sources such as fungi or nevi. Nevi can undergo modifications such as melanoma, which is a tumor generated from the melanocytes cells. This kind of tumor was investigated by D'Amico et al. (2008), who measured the volatiles emitted from the skin lesion with the LibraNose system (D'Amico et al., 2008). The study involved 40 individuals suspected of melanoma, from which 10 were finally positively diagnosed. Two samples were collected from each subject, both from the melanoma and from a region very close to the lesion, which served as a reference for the same individual and helped at counteracting intra-individual differences. To perform the experimental tests, the skin surface was first cleaned with filtered ambient air circulating in the external part of the coaxial tube of the sampler unit, and then skin headspace was taken into the sensors cell with the help of the internal pump of the electronic nose. A multivariate discriminant analysis-solved partial least squares discriminant analysis algorithm was employed to build the discriminative model between nevi and melanomas, obtaining an accuracy of 87% (90% correct classification of the benign lesions and 70% correct classification of melanoma), which is comparable with the currently employed diagnosis methods. Nevertheless, it was a much larger spread in the melanoma data as compared with the nevi data, which suggested that differences introduced by the temporal evolution of the tumor, tumor form, tumor stage and different morphological states of the examined lesions need to be taken into account.

4.2.5   ANALYSIS OF VOLATILES RELEASED BY CANCEROUS CELLS

Cell membranes can emit themselves volatile compounds, which can be analyzed directly by measuring the headspace of the cancer cells formed in a closed space above the cell lines employing the so-called in-vitro approach. This approach represents a suitable way to eliminate potential effects of confounding factors associated with the analysis of clinical samples, such as samples collection, storage and manipulation procedure, as well as differences in patients diet, age, gender or metabolic state. The direct detection of the odorant signature from cancer cells ensures that the findings are associated with the cell itself, rather than with the microen-vironment of the tumor or with indirect metabolic pathways in the body (Peled et al., 2013).

The first study employing this approach was performed by Gendron et al. (2007), who investigated the possibility to discriminate between six human tumor cell lines of the lung (four types of non-small cell lung carcinomas – L55, L65, A549, H460; metastatic squamous cell carcinoma of the lung – M51; and mesothelioma – REN), and two normal cell lines (normal human diploid fibroblasts – NHDF; and human smooth muscle – HASM) (Gendron et al., 2007). All the cell lines except HASM were grown in 10% fetal bovine serum at 37°C, 5% CO_2, while HASM was grown in Ham's F12K medium with 2 mmol/L L-glutamine and 10% fetal bovine serum at 37°C, 5% CO_2. The cells were placed in a 37°C water bath, and the headspace gasses were analyzed with Cyranose 320. Three repetitions per sample were measured to assess reproducibility. Principal Component Analysis was applied to analyze sensors data, and the Mahalanobis distances (MD) were calculated to quantify the degree of discrimination between samples. Two different studies were performed: the first one aimed to discriminate between NHDF, L55, L65 and M51, and the second one between HASM, A549, REN, and H460. The results obtained showed that the tumor cell lines were discriminated from each other, and from the normal fibroblast and smooth muscle cells. MD values ranged between 1 and 8.5, in the majority of the cases being greater than 3 (which is the threshold indicating that the fingerprints of the cell lines are distinct from each other). Several MD values were greater than 5 (e.g., L65 and M51 could be distinguished from each other and from other tumor lines with MD values between 4.36–8.47), suggesting that a trained electronic nose may be able to identify individual tumor lines, not just distinguish them from other types of cells.

Kaleb et al. (2009) used the JPL electronic nose developed for air quality monitoring at the Jet Propulsion Laboratory, California Institute of Technology, Pasadena, USA, based on an array of 16 sensors with uniquely coated polymer-carbon black composite films, to discriminate between two types of cancer cell lines: human glioblastoma – U 251, which corresponds to the most common and aggressive brain cancer; and human melanoma – A2058, which corresponds of the cancer that develops from the pigment-containing cells of the skin (Kateb et al., 2009). The cell lines were cultured in 10-mL flasks in Dulbecco's Modified Eagle's Medium. The cells were plated onto 100 mm dishes 4 days before the analysis. They were either left adherent to the culture dishes or centrifuged into pellets, then transferred into 50-mL test tubes where they were stored for

2 h at 37°C. Four preparations, each in triplicate, were totally used: 3×105 A2058 cells, 1×106 A2058 cells, 3×105 U251 cells, and 1×106 U251 cells. The headspace formed above the cultured cells was measured with the sensors array. The first result was that there was a significant difference between the medium alone and the medium with the cells, which indicated that the cells emitted VOCs that were detected by the e-nose. Regarding the influence of the number of cells used, it was found that for A2058 there was not a significant difference regarding the total variation across the sensors array between 3×105 and 1×106 cells, but the difference was significant for U251 cells (49% variance), suggesting that the number of cells can influence the results. 1×106 cells were used to analyze the differences between A2058 and U251, obtaining 22% of difference between the overall odorant patterns of the two cells. This study showed that it is possible to use an electronic nose to distinguish between two different types of tumor cells, although it is important to deeper investigate the influence of cellular proliferation, growth, differentiation and infiltration on the odorant fingerprint of different cells.

Thriumani et al. (2015) used the Cyranose 320 system to discriminate between the volatiles emitted by lung cancer cells (lung cancer cell line A549), breast cancer cells (breast cancer cell line MCF7), and the blank media (without cells) (Thriumani et al., 2015). The cancer cells were cultivated in 75 cm² T-flasks using the complete medium (90% v/v Dulbecco's Modified Eagle's Medium with 10% v/v bovine fetal serum). The concentration of each cell line was of 1×105 cells. The samples were prepared in triplicates and incubated for 72 h in a humidified atmosphere with 5% CO_2 at 37°C. The headspaces of VOCs formed in the T-flasks above the cancerous cells and blank medium were analyzed with Cyranose 320 by inserting the snout of the device into the T-flasks for 15 min. The measurement process was conducted once every 24 hours during 3 consecutive days, and each measurement was repeated three times to check reproducibility. Principal Component Analyses was used to reduce the dimensionality of the data, and the Probabilistic Neural Network was applied to investigate the performance of the e-nose system for cancerous cells discrimination. The experimental results were averaged over several runs of randomly generated 60/40 – training/test splits of the data. The results obtained proved that the e-nose could be able to detect lung cancer cells from breast cancer cells and blank media with over 99% of classification accuracy using day-1, day-2, day-3, and all three-days samples.

Peled et al. (2013) aimed the identification of representative genetic mutations in lung cancer cells that are known to be associated with targeted cancer therapy (Peled et al., 2013). Nineteen human non-small-cell lung carcinoma (NSCLC) cell lines were analyzed, including six cell lines representing the oncogene EGFRmut (H3255, H820, H1650, H1975, HCC4006, HCC2279), four cell lines representing KRASmut (A549, H2009, H460, NE18), one cell line representing EML4-ALK fusion (H2228), and seven cell lines representing oncogenes that were wild type (wt) to the three mutations of interest (H322, H1703, H125, H1435, Calu3, HCC15, H520, HCC193). The cell lines were grown in 100 mm cell culture dishes from seeding $\sim$ 2–7 × 10^6 cells in a conventional incubator at 37°C in a humidified atmosphere with 95% air/5% CO_2, using a two-dimensional medium (Roswell Park Memorial Institute 1640 medium + 10% fetal bovine serum). Two badges with Tenax TA sorbent material were placed above the cell-culture, attached to the cover of the dish, for absorbing the headspace VOCs during the total growth time (median time 68 h; range 60–72 h). The cell lines were grown in several replicates, while an empty medium with the same incubation time and conditions, but without cells, served as control. The headspace samples from the cell lines were analyzed with the NA-NOSE system, and Discriminant Function Analysis was employed to build classification models. Leave-one-out cross-validation yielded a classification success of 70% sensitivity, 100% specificity and 92% accuracy for EGFRmut vs. EGFTwt; 93% sensitivity and 78% specificity for KRASmut vs. KRASwt; and 63% sensitivity and 100% specificity for EML4-ALK. While these results were obtained through in-vitro studies, the authors consider that it is reasonable to assume that similar odorant patterns could be detected directly from blood or exhaled breath samples, which would allow immediate testing to predict the genetic mutation and to prescribe targeted therapy.

4.3 OUTLOOK AND CONCLUSIONS

The identification of cancer-specific odorant signatures in body fluid samples employing artificial olfactory systems is a novel approach that is still at the equivalent level of its initial infancy.

Two high-level research groups are focusing considerable efforts on developing and testing e-nose devices in this field, formed of own specifically designed and fabricated gas sensors: the group of Prof. Haick from Technion – Israel Institute of Technology, and the group of Prof. Di Natale from University of Rome Tor Vergata, Italy. Prof. Haick developed and tested a nanomaterials-based array of chemiresistive sensors for the diagnosis of a wide range of diseases through exhaled breath analysis, while Prof. Di Natale developed an array of QCM sensors coated with different metalloporphyrins and tested it for the diagnosis of various cancers through the analysis of the volatiles released by different body fluids. These systems were referred as "NA-NOSE" and "LibraNose," respectively, within this chapter, although they were not always mentioned with these names in the scientific reports published by the mentioned researchers.

Other studies were performed with arrays of gas sensors fabricated by the research groups that analyzed the samples, or with commercially available e-nose systems such as for instance Cyranose® 320 or AEONOSETM (former DiagNose). However, these artificially olfactory systems were not specifically designed for biomedical applications like the one presented in this chapter. For instance, Cyranose 320 is a vapor sensing instrument designed to detect and identify complex chemical mixtures that constitute aromas, odors, fragrances, formulations, spills and leaks, and the fabricant says that it can be used in diverse industries including petrochemical, chemical, food and beverage, packaging materials, plastics, pet food, pulp and paper, medical research, and many more. Nevertheless, the NA-NOSE was designed to target the detection of non-polar compounds under humidity conditions reproducing the moisture content present in most of the body fluid samples; the non-polar compounds serve as diagnostic markers of various diseases and are generally more difficult to detected than the polar ones – for instance, Cyranose 320 cannot detect them (Bikov et al., 2015). Although it initially comprised monolayers of organically capped gold nanoparticles, posteriorly other types of sensors that complied with this condition were added to the NA-NOSE for enhancing the number and classes of compounds detected, such as synthetically designed polycyclic aromatic hydrocarbons derivatives with different aromatic coronae and side groups and hexa-peri-hexabenzocoronene derivatives with different functional hydrophobic side groups (Zilberman et al., 2009, 2011).

As already mentioned in this chapter, the sensors that form an artificial olfactory system are not specific to a certain compound, but rather to the overall signature of the odor. Therefore, although the analytical chemistry techniques give a clue about the biomarkers whose detection should be targeted, the sensors array is not fabricated with sensors that can specifically and uniquely detect these biomarkers. Instead, the strategy that is employed in diseases diagnosis through the measurement of body fluid samples with artificially olfactory systems is based on measuring the samples with a reservoir of many sensors, and then heuristically select the optimal sensors for the foreseen application based on sensors responses, using the pattern recognition algorithm as selection approach (Tisch et al., 2013). Obviously, the particular sensors selected depend on the disease targeted (Xu et al., 2013).

In view of these considerations, it is obvious that the electronic nose technology didn't achieve the maturity yet in order to can replace the current diagnostic methods employed in clinical use. But, supported by the possibility to measure non-invasive samples and by the low cost of such a measurement [~ €10 (Rocco et al., 2015)], the e-nose could be very useful as a mass screening tool for identifying at-risk individuals that should undergo further investigations. The results of such a test could potentially provide valuable complementary information and could detect the disease at an early stage as metabolomic changes occur in the organism with the very onset of the disease. They could also identify malignant lesions missed by the current standard diagnostic techniques, or with a similar appearance with benign lesions.

Finally, it is worth pointing out that, although of the very promising results reported till now, the studies performed so far were realized on small-size study groups that did not overstep the proof-of-concept. Large-scale clinical studies conceived and realized together with medical professionals is surely needed to validate these initial results.

On the other hand, VOCs are emitted by a multitude of body fluids: cells, breath, sputum, saliva, skin (sweat), blood (plasma, serum), urine, feces, vaginal secretion, and ear wax, but only a small number of VOCs (around 1%) were found in the main human body sources (breath, skin, urine, blood, saliva, and feces) (Broza et al., 2015). However, only several body fluid samples that were highlighted in this chapter (breath, urine, feces, skin, and cells) were analyzed with artificial olfactory systems for cancer detection. Therefore, there is a huge uncovered place in this research field that undoubtedly worth to capture the attention and effort of the scientific community. Moreover, the concomitant study of several body fluids (referred as "hybrid volatolomics") (Broza et al., 2015) could provide complementary information rather than duplicating it, and hence improve the diagnostic results; this approach was not assessed so far for cancer diagnosis with artificial olfactory systems.

KEYWORDS

- artificial olfactory systems
- body fluids
- cancer
- exhaled breath
- LibraNose
- NA-NOSE

REFERENCES

Adiguzel, Y., & Kulah, H., (2015). Breath sensors for lung cancer diagnosis. *Biosens. Bioelectron.*, *65*, 121–138.

Amal, H., Leja, M., Funka, K., Skapars, R., Sivins, A., Ancans, G., Liepniece-Karele, I., Kikuste, I., Lasina, I., & Haick, H., (2016). Detection of precancerous gastric lesions and gastric cancer through exhaled breath. *Gut.*, *65*(3), 400–407.

Amal, H., Shi, D. Y., Ionescu, R., Zhang, W., Hua, Q. L., Pan, Y. Y., Tao, L., Liu, H., & Haick, H., (2015). Assessment of ovarian cancer conditions from exhaled breath. *Int. J. Cancer.*, *136*(6), 614–622.

Arasaradnam, R. P., Ouaret, N., Thomas, M., Gold, P., Quraishi, M., Nwokolo, C. U., Bardhan, K. D., & Covington, J. A., (2012). Evaluation of gut bacterial populations using an electronic e-nose and field asymmetric ion mobility spectrometry: further insights into 'fermentonomics.' *J. Med. Eng. Technol.*, *36*(7), 333–337.

Arshak, K., Moore, E., Lyons, G., Harris, J., & Clifford, S., (2004). A review of gas sensors employed in electronic nose applications. *Sensor Rev.*, *24*(2), 181–198.

Asimakopoulos, A., Del Fabbro, D., Miano, R., Santonico, M., Capuano, R., Pennazza, G., D'Amico, A., & Finazzi-Agro, E., (2014). Prostate cancer diagnosis through electronic nose in the urine headspace setting: a pilot study. *Prostate Cancer Prostatic Dis.*, *17*(2), 206–211.

Bernabei, M., Pennazza, G., Santonico, M., Corsi, C., Roscioni, C., Paolesse, R., Di Natale, C., & D'Amico, A., (2008). A preliminary study on the possibility to diagnose urinary tract cancers by an electronic nose. *Sens. Actuator. B-Chem.*, *131*(1), 1–4.

Bikov, A., Lázár, Z., & Horvath, I., (2015). Established methodological issues in electronic nose research: how far are we from using these instruments in clinical settings of breath analysis? *J. Breath Res.*, *9*(3), 034001.

Bishop, C. M., (2006). *Pattern Recognition and Machine Learning, 1st edn.*, ISBN: 978-0-387-31073-2, Springer-Verlag: New York, pp. 738.

Broza, Y. Y., Kremer, R., Tisch, U., Gevorkyan, A., Shiban, A., Best, L. A., & Haick, H., (2013). A nanomaterial-based breath test for short-term follow-up after lung tumor resection. *Nanomedicine, 9*(1), 15–21.

Broza, Y. Y., Mochalski, P., Ruzsanyi, V., Amann, A., & Haick, H., (2015). Hybrid volatolomics and disease detection. *Angew. Chem. Int. Ed. Engl., 54*(38), 11036–11048.

D'Amico, A., Bono, R., Pennazza, G., Santonico, M., Mantini, G., Bernabei, M., Zarlenga, M., Roscioni, C., Martinelli, E., Paolesse, R., & Di Natale, C., (2008). Identification of melanoma with a gas sensor array. *Skin Res. Technol., 14*(2), 226–236.

De Vries, R., Brinkman, P., Van der Schee, M., Fens, N., Dijkers, E., Bootsma, S., de Jongh, F., & Sterk, P., (2015). Integration of electronic nose technology with spirometry: validation of a new approach for exhaled breath analysis. *J. Breath Res., 9*(4), 046001.

Di Natale, C., Macagnano, A., Martinelli, E., Paolesse, R., D'Arcangelo, G., Roscioni, C., Finazzi-Agrò, A., & D'Amico, A., (2003). Lung cancer identification by the analysis of breath by means of an array of non-selective gas sensors. *Biosens. Bioelectron., 18*(10), 1209–1218.

Di Natale, C., Paolesse, R., Martinelli, E., & Capuano, R., (2014). Solid-state gas sensors for breath analysis: a review. *Anal. Chim. Acta., 824*, 1–17.

Dragonieri, S., Van der Schee, M. P., Massaro, T., Schiavulli, N., Brinkman, P., Pinca, A., Carratú, P., Spanevello, A., Resta, O., & Musti, M., (2012). An electronic nose distinguishes exhaled breath of patients with malignant pleural mesothelioma from controls. *Lung Cancer, 75*(3), 326–331.

Gardner, J. W., & Bartlett, P. N., (1994). A brief history of electronic noses. *Sens. Actuator. B-Chem., 18*(1–3), 210–211.

Gasparri, R., Santonico, M., Valentini, C., Sedda, G., Borri, A., Petrella, F., Maisonneuve, P., Pennazza, G., D'Amico, A., & Di Natale, C., (2016). Volatile signature for the early diagnosis of lung cancer. *J. Breath Res., 10*(1), 016007.

Gendron, K. B., Hockstein, N. G., Thaler, E. R., Vachani, A., & Hanson, C. W., (2007). In vitro discrimination of tumor cell lines with an electronic nose. *Otolaryngol. Head Neck Surg., 137*(2), 269–273.

Gruber, M., Tisch, U., Jeries, R., Amal, H., Hakim, M., Ronen, O., Marshak, T., Zimmerman, D., Israel, O., & Amiga, E., (2014). Analysis of exhaled breath for diagnosing head and neck squamous cell carcinoma: a feasibility study. *Br. J. Cancer, 111*(4), 790–798.

Hakim, M., Billan, S., Tisch, U., Peng, G., Dvrokind, I., Abdah-Bortnyak, R., Kuten, A., & Haick, H., (2011). Diagnosis of head-and-neck-cancer from exhaled breath. *Br. J. Cancer., 104*, 1649–1655.

Hooren, M. R., Leunis, N., Brandsma, D. S., Dingemans, A.-M. C., Kremer, B., & Kross, K. W., (2016). Differentiating head and neck carcinoma from lung carcinoma with an electronic nose: a proof of concept study. *Eur. Arch. Otorhinolaryngol.*, 1–7.

Kateb, B., Ryan, M., Homer, M., Lara, L., Yin, Y., Higa, K., & Chen, M. Y., (2009). Sniffing out cancer using the JPL electronic nose: A pilot study of a novel approach to detection and differentiation of brain cancer. *Neuroimage, 47*, T5–T9.

Krilaviciute, A., Heiss, J. A., Leja, M., Kupcinskas, J., Haick, H., & Brenner, H., (2015). Detection of cancer through exhaled breath: a systematic review. *Oncotarget, 6*(36), 38643–38657.

Leunis, N., Boumans, M. L., Kremer, B., Din, S., Stobberingh, E., Kessels, A. G., & Kross, K. W., (2014). Application of an electronic nose in the diagnosis of head and neck cancer. *Laryngoscope*, *124*(6), 1377–1381.

Meij, T. G., Larbi, I. B., Schee, M. P., Lentferink, Y. E., Paff, T., Terhaar sive Droste, J. S., Mulder, C. J., Bodegraven, A. A., & Boer, N. K., (2014). Electronic nose can discriminate colorectal carcinoma and advanced adenomas by fecal volatile biomarker analysis: proof of principle study. *Int. J. Cancer*, *134*(5), 1132–1138.

Narayanan, V., Peppelenbosch, M. P., & Konstantinov, S. R., (2014). Human fecal microbiome–based biomarkers for colorectal cancer. *Cancer Prev. Res.*, *7*(11), 1108–1111.

Nardi-Agmon, I., Abud-Hawa, M., Liran, O., Gai-Mor, N., Ilouze, M., Onn, A., Bar, J., Shlomi, D., Haick, H., & Peled, N., (2016). Exhaled breath analysis for monitoring response to treatment in advanced lung cancer. *J. Thorac. Oncol.*, *11*(6), 827–837.

Peled, N., Barash, O., Tisch, U., Ionescu, R., Broza, Y. Y., Ilouze, M., Mattei, J., Bunn, P. A., Hirsch, F. R., & Haick, H., (2013). Volatile fingerprints of cancer specific genetic mutations. *Nanomedicine*, *9*(6), 758–766.

Peled, N., Hakim, M., Bunn, P. A. J., Miller, Y. E., Kennedy, T. C., Mattei, J., Mitchell, J. D., Hirsch, F. R., & Haick, H., (2012). Non-invasive breath analysis of pulmonary nodules. *J. Thorac. Oncol.*, *7*(10), 1528–1533.

Peng, G., Hakim, M., Broza, Y. Y., Billan, S., Abdah-Bortnyak, R., Kuten, A., Tisch, U., & Haick, H., (2010). Detection of lung, breast, colorectal, and prostate cancers from exhaled breath using a single array of nanosensors. *Br. J. Cancer*, *103*(4), 542–551.

Rocco, R., Incalzi, R. A., Pennazza, G., Santonico, M., Pedone, C., Bartoli, I. R., Vernile, C., Mangiameli, G., La Rocca, A., & De Luca, G., (2015). BIONOTE e-nose technology may reduce false positives in lung cancer screening programmes. *Eur. J. Cardiothorac. Surg.*, *49*(4), pp. 1112–1117, https://doi.org/10.1093/ejcts/ezv328.

Rock, F., Barsan, N., & Weimar, U., (2008). Electronic nose: Current status and future trends. *Chem. Rev.*, *108*, 705–725.

Roine, A., Veskimäe, E., Tuokko, A., Kumpulainen, P., Koskimäki, J., Keinänen, T. A., Häkkinen, M. R., Vepsäläinen, J., Paavonen, T., & Lekkala, J., (2014). Detection of prostate cancer by an electronic nose: a proof of principle study. *J. Urol.*, *192*(1), 230–235.

Santonico, M., Pennazza, G., Grasso, S., D'Amico, A., & Bizzarri, M., (2013). Design and test of a biosensor-based multisensorial system: A proof of concept study. *Sensors*, *13*(12), 16625–16640.

Shirasu, M., & Touhara, K., (2011). The scent of disease: volatile organic compounds of the human body related to disease and disorder. *J. Biochem.*, *150*(3), 257–266.

Shuster, G., Gallimidi, Z., Reiss, A. H., Dovgolevsky, E., Billan, S., Abdah-Bortnyak, R., Kuten, A., Engel, A., Shiban, A., & Tisch, U., (2011). Classification of breast cancer precursors through exhaled breath. *Breast Cancer Res. Treat.*, *126*(3), 791–796.

Thriumani, R., Jeffreea, A. I., Zakaria, A., Hasyim, Y. Z. H.-Y., Helmy, K. M., Omar, M. I., Adom, A. H., Shakaff, A. Y., & Kamarudin, L. M., (2015). A preliminary study on detection of lung cancer cells based on volatile organic compounds sensing using electronic nose. *J. Teknol.*, *77*(7), 67–71.

Tisch, U., Schlesinger, I., Ionescu, R., Nassar, M., Axelrod, N., Robertman, D., Tessler, Y., Azar, F., Marmur, A., & Aharon-Peretz, J., (2013). Detection of Alzheimer's and Parkinson's disease from exhaled breath using nanomaterial-based sensors. *Nanomedicine*, *8*(1), 43–56.

Westenbrink, E., Arasaradnam, R. P., O'Connell, N., Bailey, C., Nwokolo, C., Bardhan, K. D., & Covington, J., (2015). Development and application of a new electronic nose instrument for the detection of colorectal cancer. *Biosens. Bioelectron.*, *67*, 733–738.

Xu, Z. Q., Broza, Y. Y., Ionsecu, R., Tisch, U., Ding, L., Liu, H., Song, Q., Pan, Y. Y., Xiong, F. X., Gu, K. S., Sun, G. P., Chen, Z. D., Leja, M., & Haick, H., (2013). A nanomaterial-based breath test for distinguishing gastric cancer from benign gastric conditions. *Br. J. Cancer*, *108*(4), 941–950.

Zilberman, Y., Ionescu, R., Feng, X., Mullen, K., & Haick, H., (2011). Nanoarray of polycyclic aromatic hydrocarbons and carbon nanotubes for accurate and predictive detection in real-world environmental humidity. *ACS Nano.*, *5*, 6743–6753.

Zilberman, Y., Tisch, U., Pisula, W., Feng, X., Müllen, K., & Haick, H., (2009). Sponge-like structures of hexa-peri-hexabenzocoronene derivatives enhances the sensitivity of chemiresistive carbon nanotubes to nonpolar volatile organic compounds. *Langmuir*, *25*(9), 5411–5416.

CHAPTER 5

BOTTOM-UP CELL CULTURE MODELS TO ELUCIDATE HUMAN *IN VITRO* BIOMARKERS OF INFECTION

MICHAEL SCHIVO,[1,2] MITCHELL M. MCCARTNEY,[3]
MEI S. YAMAGUCHI,[3] EVA BORRAS,[3] and CRISTINA E. DAVIS[3]

[1]*Department of Medicine, Division of Pulmonary, Critical Care, and Sleep Medicine, University of California, Davis, Sacramento, CA, USA*

[2]*Center for Comparative Respiratory Biology and Medicine, University of California, Davis, Davis, CA, USA*

[3]*Department of Mechanical and Aerospace Engineering, University of California, Davis, One Shields Avenue, Davis, CA, 95616, USA, E-mail: cedavis@ucdavis.edu*

ABSTRACT

Respiratory bacterial, fungal, and viral infections can lead to substantial morbidity and mortality in people with underlying lung diseases, but may also cause problems for healthy adults. Although some rapid tests are available, they lack accuracy to be widely helpful and rely on direct pathogen identification. However, it appears that upper respiratory infections (URIs) may be non-invasively diagnosed by analyzing exhaled breath for specific host-response-to-infection metabolites along with metabolites from microorganisms themselves. Exhaled breath contains thousands of volatile organic compounds (VOCs) and inflammatory mediated markers. While past phenomenological studies have associated specific breath chemicals to specific diseases, there is no mechanistic framework in place to predict or understand how these biomarkers relate to overall health status. We propose that upon infection, species- and strain-specific cascades of intracellular signaling mechanisms are initiated that have metabolic end points released into exhaled human breath. By examining host-response to infection and metabolites generated from microorganisms themselves, we can build a "bottom-up" model of breath metabolites to understand better how they are generated and released in the human body.

5.1   INTRODUCTION TO BREATH ANALYSIS

The analysis of small, lightweight metabolites in human breath is a relatively new and exciting field (Davis et al., 2010b). In truth, physicians and laypersons alike have used human breath to detect conditions such as liver disease and anaerobic oral and gut infections for centuries. However, the systematic study of human breath, including the study of metabolites from cell culture models that may ultimately be seen in human breath, is a recent advent. Breath contains both VOC and non-VOC (Davis et al., 2010a) that, together, may contain useful information about a person's health status. In fact, over 800 VOC species have been described in breath to date (Filipiak et al., 2016; Sobus, 2013), although most have not yet been mapped to relevant metabolic pathways. As technology improves, cell culture and human breath data become more available, and an improved understanding of metabolite origin develops, breath analysis is likely to become a major diagnostic platform.

Breath analysis can be divided into three main parts: breath collection, metabolite detection, and data analysis. Each part requires careful attention to principles of standardization and reproducibility, and at present, the breath field is focusing on validated methods. Breath collection techniques require attention to problems of contamination, methods to ensure specific sets of metabolites are captured, and, in human studies, acceptable collection maneuvers. Metabolite detection depends on specific analytic instrumentation such as mass spectrometry (MS). However, the range of VOC and non-VOC metabolites is broad, and no single detection method has yet been able to capture all chemical species. Some instrumentation progress has been made (Bumatay et al., 2012; Cumeras et al., 2015a, b; Kwan et al., 2015; Strand et al., 2010; Zrodnikov et al., 2013), but there is ample room to advance. Moreover, data analysis techniques vary widely, though these often involve data visualization, clustering, and data modeling techniques (Sobus, 2013).

Human breath is complex and comprised of VOCs and non-VOCs from several different cell types and organs (Amann et al., 2014), and there is growing attention to cell culture models to simplify and compartmentalize the origins of metabolites (Aksenov et al., 2013). This applies to both infectious and non-infectious models. The ability to pair metabolite data from cell culture with data from clinical breath studies will hopefully lead to a deeper understanding of metabolite origin that translates to meaningful diagnostic outcomes.

In this chapter, we present how metabolites can be used to identify a range of infections in humans by utilizing cell culture models. By understanding how cell culture models can elucidate a "library" of both VOC and non-VOC metabolites, one can begin to uncover specific components of human breath, which may ultimately be used to identify early-stage infections before symptoms appear.

5.2   CONCEPT OF BREATH METABOLOMICS

5.2.1   BREATH METABOLITES EXIST ACROSS A RANGE OF VOLATILITY

Human breath is known to contain trace amounts of VOCs such as acetone, methanol, ethanol, aldehydes, and alkanes, typically in the range of parts-per-billion (ppb) or parts-per-trillion (ppt) molar fraction (Aksenov et al., 2013). Nonvolatile compounds such as lipids, proteins/peptides, virus particles or entire bacteria or epithelial cells may also be present in breath aerosol droplets carried in the exhaled breath stream (Aksenov et al., 2014b). Breath aerosols are a mixture of water and metabolite content derived from the lung lining fluid, and, as such, they may partially reflect the metabolic profile of blood (Aksenov et al., 2014b). Hundreds of different endogenous and exogenous chemicals have been identified in the exhaled breath and other biogenic fluids of humans (Amann et al., 2014; Filipiak et al., 2016).

Breath analysis is not only an attractive assessment tool for humans but also for animal models due to its noninvasive nature. For example, dolphin breath contains a large variety of low-abundance metabolites, many of which are also found in human breath (Aksenov et al., 2014b). The noninvasive method of breath analysis may provide a very valuable tool in future wildlife conservation efforts as well as deepen our understanding of wildlife biology and physiology.

5.2.2 BREATH BIOMARKER PANELS CHANGE WITH PHYSIOLOGY

Exhaled breath is a valuable medium to study how metabolites change with physiology. Previous studies have shown that the breath biomarkers panel will change with disease or its physiology; moreover, different biomarkers can be detected in different disease or physiology in breath. Figueroa et al. used GC-MS to study exhaled breath condensate from subjects before and after being exposed to normobaric hypoxia. Some patients had a history of high altitude pulmonary edema (HAPE). Two compounds (benzyl alcohol and dimethylbenzaldehyde) found in the breath samples could distinguish between those with and without a history of HAPE. This suggests that key metabolic differences exist in patients who have been afflicted by HAPE. Ketones, aromatic compounds, and monoterpenes were uniquely found in idiopathic pulmonary arterial hypertension (IPAH) patients exhaled breath (Mansoor et al., 2014). Nording et al. found five fatty acid metabolites: 9,12,13- trihydroxyoctadecenoic acid (9,12,13-TriHOME), 9,10,13-TriHOME, 12,13- dihydroxyoctadecenoic acid (12,13-DiHOME), 12-hydroxyeicosatetraenoic acid (12-HETE), and 12 (13)-epoxyoctadecenoic acid (12(13)-EpOME) in Exhaled Breath Condensate (EBC) that may be potential biomarkers for asthma monitoring and diagnosis (Nording et al., 2010).

5.2.3 DETECTION OF VOCs

Due to the complex chemical compositions of both exhaled breath and cell culture models, there is no single instrumental technique able to detect all constituent metabolites. To measure a relatively wide range of compounds simultaneously the most common techniques are gas and liquid chromatography (GC and LC, respectively) coupled with MS (Peralbo-Molina et al., 2015, 2016; van der Schee et al., 2014) and nuclear magnetic resonance (NMR) (Sofia et al., 2011). These analytical platforms generate massive amounts of metabolic data and can be exploited to distinguish health conditions. However, numerous processing and statistical tools are required, including pre-processing, metabolite identification and statistical analysis (Nagana Gowda et al., 2008, van der Greef and Smilde, 2005; Yi et al., 2016). Here we present a generalized process flow to outline this concept (Figure 5.1).

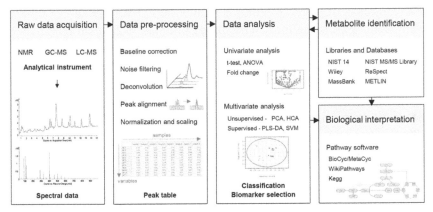

FIGURE 5.1 **(See color insert.)** Typical steps needed for breath biomarker discovery.

Data pre-processing improves signal quality by removing bias and variance. Inherently every sensor platform will contribute some amount of noise or error into their signals. Methods such as deconvolution, alignment, noise filtering, baseline correction, and normalization can reduce instrument error. Furthermore, analytical signals may contain overlapped metabolites in one single signal. Deconvolution procedures are required to tease apart constituent compounds from one signal. Alignment methods rectify non-linear shifts (e.g., time retention) that can affect all samples. Instrument error also increases the variance on measured metabolite abundances and can be reduced by proper normalization and scaling techniques (Alonso et al., 2015; Hendriks et al., 2011; Peirano et al., 2013).

The pre-processed data can easily be treated by metabolite-identification tools. While both MS and NMR provide clues about the chemical structure of each metabolite, reliable identification is challenging due to the biochemical diversity of the breath metabolites. Many metabolites may be different structural arrangements of the same chemical formula.

Distinguishing these isomers can prove challenging. Often, researchers use well-established databases and libraries as well as the use of MS platforms to improve structure elucidations and ultimately identify breath metabolites (Michael et al., 2011).

Some univariate statistical tools are commonly applied, such as the Student's t-test and analysis of variance (ANOVA), to assess differences between two or more sample classifications. However, disease diagnostics often requires correlation of multiple metabolites. Thus, the use of multivariate techniques is the most widespread practice. Statistical techniques can discover patterns in large sets of data either regardless of any sample classification (unsupervised) or with regards to sample classification, such as "healthy" and "infected" (supervised). Unsupervised methods, such as principal component analysis (PCA), hierarchical cluster analysis (HCA), and self-organization mapping (SOM) are applied to find clusters or groups within the data and enable visualization of the sample distribution. Supervised methods, such as partial least-squares discriminant analysis (PLS-DA), linear discriminant analysis (LDA), orthogonal projections to latent structures-discriminant analysis (OPLS-DA) or support vector machine (SVM), use previously known data information (such as infection status) to look for differences between sample sets and can develop statistical models to predict the classification (health status) of new samples (Alonso et al., 2015; Hendriks et al., 2011).

5.3 BACTERIAL CELL CULTURE MODELS OF RESPIRATORY INFECTION

Bacterial infections are a major threat to lung health, routinely affecting people with chronic lung diseases such as cystic fibrosis. Sputum cultures in cystic fibrosis patients are often positive for bacteria, though differentiating between bacterial colonization versus infection can be difficult, leading to potentially inappropriate antibiotic use (Mogayzel et al., 2013). Thus, there is a need to understand the VOCs that originate from bacteria and bacteria-human interactions to inform breath diagnostic studies. This may translate to improved and non-invasive diagnostic platforms that do not rely on sputum cultures.

5.3.1   BACTERIA EMIT VOLATILE ORGANIC COMPOUNDS (VOCs)

Studies have indicated that specific pathogenic bacteria emit VOCs that can be captured and measured (Chen et al., 2016; Shnayderman et al., 2005). Bacteria are known to emit a broad range of VOCs, such as hydrocarbons (Junger et al., 2012; Tait et al., 2014b), aliphatic alcohols (Junger et al., 2012; Tait et al., 2014b), ketones (Bhaumik et al., 2005), aromatic compounds (Humphris et al., 2002; Robacker and Lauzon, 2002), sulfur compounds (Junger et al., 2012), and terpenoids (Schulz and Dickschat, 2007; Yamada et al., 2015). VOCs are likely by-products of metabolic pathways, including the generation of hydrocarbons from fatty acid synthesis and indole from amino acid catabolism (Schulz and Dickschat, 2007; Tait et al., 2014b). Furthermore, bacteria may emit different VOCs based on their growth phase (Shnayderman et al., 2005), suggesting an opportunity to identify bacteria when they are causing an infection rather than colonization.

5.3.2   BACTERIA VOCs FOR PULMONARY INFECTION DIAGNOSTICS

There is major interest in identifying bacterial VOC signatures that can ultimately be used in breath diagnostics. Cystic fibrosis is a good clinical model of chronic bacterial infection, and culture studies have focused on the known bacteria that often infect cystic fibrosis patients (Dryahina et al., 2016; Neerincx et al., 2016). These bacteria include *Staphylococcus aureus*, *Pseudomonas aeruginosa*, *Burkholdaria* spp., *Haemophilus* spp, and *Stenotrophamonas* spp. Dryahina and colleagues assessed VOCs from several bacteria using selected ion-flow tube mass spectrometry (SIFT-MS) (Dryahina et al., 2016). They showed that among a large overlap of compounds there were specific VOCs for different bacteria. For example, hydrogen cyanide and methyl thiocyanate were specific for *Pseudomonas aeruginosa*; isoprene for *Burholdaria cepacia*; and butyric acid for *Staphylococcus aureus*. Bos et al. (2016) found that ethanol, acetaldehyde, and methanol in human breath were predictive of bacterial infections. They recovered *Escherichia* and *Pseudomonas* spp. from cystic fibrosis patients and conducted metagenomics sequencing. The bacteria contained the required genetic information to produce the alcohols and aldehydes seen in infected breath samples.

However, human breath is biologically complex and influenced by multiple factors (e.g., diet, gut/oral microbiome, inflammation, etc.). One must be cautious to translate cell culture findings directly to breath studies. This was demonstrated by a study of 58 cystic fibrosis patients where their breath contained significantly higher levels of acetic acid compared to healthy controls, independent of infection (Smith et al., 2016); however, it's unclear the exact biological origin of specific VOCs.

5.4   FUNGUS CELL CULTURE MODELS OF VOLATILE PRODUCTION

Fungi also emit VOCs, but fungal VOCs have not been well studied compared to other organisms. Many studies focus on culinary applications, especially valuable truffles (Costa et al., 2015; Vahdatzadeh et al., 2015), but plant-fungus and insect-fungus interactions via volatile emissions have also been of interest (Li et al., 2016). Breath-based pulmonary diagnostics could also benefit from fungal emission studies, either from studies that look for unique fungal VOCs or human byproducts of fungal infections.

5.4.1   FUNGI EMIT VOCs DURING PROLIFERATION

Some 300 fungal volatiles have been identified, and a database has been established to archive these compounds (Lemfack et al., 2014). Fungi have been reported to emit alcohols, benzenoids, aldehydes, alkenes, acids, esters, terpenoids, ketones and other chemical classes. Like other living organisms, fungi can communicate through VOCs. One of the most commonly produced fungal volatiles, 1-octen-3-ol, was found to have hormonal effects on fungal development (Chitarra et al., 2004). Nemčovič et al. found that conidiating fungal colonies off-gassed eight-carbon compounds, such as 1-octen-3-ol. They could induce conidiation of another fungal colony by introducing 1-octen-3-ol vapor (Nemcovic et al., 2008). Sesquiterpenes help fungi defend against predators and can attract pollinators to spread spores (Kramer and Abraham, 2012). Thus, diagnostic tools can take advantage of these emissions for pulmonary disease detection.

5.4.2   FUNGI VOCs FOR PULMONARY DIAGNOSTICS

Fungal respiratory disorders often occur from inhaling airborne spores, especially during building demolitions or handling moist soil or decomposing matter. Aspergillosis, histoplasmosis, fungal pneumonia, and blastomycosis are examples of pulmonary infections by fungi. Some have symptoms that mimic non-fungal infections. For diagnostics, focus has been spent on identifying unique volatile biomarkers specific to fungal infections. Chambers et al. measured breath from healthy controls and patients infected with *Aspergillus fumigatus* using GC-MS. They detected 2-pentylfuran in breath from subjects with *Aspergillus* infections and could not associate 2-pentylfuran with healthy subjects (Chambers et al., 2009). Furthermore, breath tests for fungal infections could utilize volatile compounds only emitted by a certain type or species of fungus. This was proven plausible, as a study by Matysik et al. found that some cultured fungal species produce unique VOC signals. For example, *Aspergillus fumigatus* emitted butoxyethoxyethanol and 2-nonen-1-ol, which were not observed in five other types of fungi (Matysik et al., 2008). These unique VOC signatures could help establish breath-based volatile tests for specific fungal pulmonary diseases, but this area requires more effort in understanding fungal-pulmonary interactions and their effects on breath emissions.

5.5   VIRAL RESPIRATORY INFECTION CHANGES VOCs

As with any microbe-human cell interaction, viruses that infect human airway cells engender several changes, including an alteration of how human genes are regulated (Troy and Bosco, 2016). The link between these genetic changes, inflammation, and the subsequent VOC release is highly plausible yet currently unproven. In addition to infected cells, uninfected but activated inflammatory cells likely lead to many of the VOCs seen in human breath during an infection (Aksenov et al., 2013). This may include VOCs seen from recruited neutrophils, macrophages and lymphocytes.

5.5.1 MAMMALIAN CELLS RELEASE VOCS

The complexity of metabolites observed in human breath likely reflects the array of cells that produce VOCs. Thus, looking toward *in vitro* models of VOC production from single or a limited number of cells in isolation is attractive. This approach may enable the identification of compounds that are uniquely produced by only a few cell types and may improve specificity. It has been described that a variety of mammalian cells release VOCs in culture. Examples include human airway epithelial cells (Schivo et al., 2014), cancer cells (Filipiak et al., 2010, 2016), neutrophils (Bansal et al., 2012), and lymphocytes (Aksenov et al., 2012) among others. For example, Schivo et al. described the production of several metabolites representing nonspecific cellular oxidation products over undisturbed bronchial epithelial cells (Schivo et al., 2014), and Aksenov et al. described the baseline production of metabolites from cultured human lymphoblastoid cells (Aksenov et al., 2012). Most of the VOCs identified cannot be reliably traced back to a specific cell type since most *in vitro* studies were designed as an untargeted identification of headspace VOCs. However, the production of some VOC molecules, such as exhaled nitric oxide and VOCs from some cancer cells, have been mapped to specific enzyme pathways (Haick et al., 2014; Vanhoutte, 2013).

5.5.2 MAMMALIAN CELL CULTURE MODEL OF GENE EXPRESSION DIFFERENCES

Though there are several pathways likely involved in VOC production, VOCs that can be used to distinguish between cell types (or cell states, e.g., infected or uninfected) likely vary based on allelic differences. In a study of cultured B cells, Aksenov et al. (2012) showed that single gene differences reproducibly change the VOC profiles seen. B cells were transfected with different human leukocyte antigen (HLA) genes, which code for a variety of downstream metabolic products. In a comparison of around 14 distinct compounds, both metabolite differences and abundance differences reproducibly distinguished between 6 cell lines. Though the compounds and the biochemical pathways of origin were not identified, this was one of the first studies to show that gene expression differences lead to a change of VOC profiles. As HLA genes are involved with host immune response, VOC profiling of HLA gene activation has important implications for the detection of cancer, infection, and non-infectious inflammation. Peled et al. (2013) studied cultured cancer cells with genetic mutations known to be associated with specific, targeted therapies (e.g., EML4-ALK, EGFR*mut*, and KRAS*mut*). Though the study was small, specific VOC patterns of five metabolites (e.g., trimethylamine, toluene, styrene, benzaldehyde, and decanal) distinguished between cell types. The use of VOCs for cellular genetic differences could ultimately lead to more specific therapies that obviate the need for tissue biopsy.

5.5.3   VIRAL CELL CULTURE MODELS FOR RESPIRATORY INFECTION

Several researchers have measured and compared the VOCs emitted by healthy and virus-infected mammalian cells grown in standard cell culture media. Aksenov et al. (2014a) took human B-lymphoblastoid cells and infected a subset with one of three influenza viruses: H9N2, H6N2, and H1N1. Using solid phase microextraction (SPME)-GC-MS, they measured the volatiles from the four cell treatments. Infection-specific VOCs did not appear in healthy control cells and only in infected cells, such as (putatively) 2-methoxy-ethanol. Other volatiles changed with time post-infection. For example, some compounds were present 24 h after infection but disappeared at a 48-h measurement. Other compounds were found at different abundances between the two-time points. In another cell culture example, Schivo et al. (2014) measured volatiles from healthy and human rhinovirus infected human primary tracheobronchial epithelial cells. Their SPME-GC-MS method found 16 volatiles differentially expressed between the healthy and HRV infected epithelial cells, depending on the time post infection. One biomarker, dimethyl sulfide, is likely due to the HRV virus causing host cell proteolysis.

These models demonstrate that viral infections alter the volatile emissions of mammalian cells. Furthermore, altered VOC emissions change with the course of viral infection, likely due to the cascade of biological events that occur within the cells. These biomarkers may permit the development of breath-based diagnostic tools to detect viral infections or even provide information to the length of infection.

5.6   CONCLUSIONS AND FUTURE TRENDS

Clearly, the field of non-invasive breath diagnostics is undergoing a rapid transition from a nascent field of inquiry into a more developed field of research that goes beyond simple phenomenology. Early studies in the field focused on clinical trials and untargeted biomarker discovery, but this approach is sensitive to artifacts and confounding effects in the clinical study design. Due to the untargeted nature of this biomarker search paradigm, these artifacts are very difficult to identify. Extremely large clinical trials with high numbers of human subjects are needed to fully verify these biomarker identifications via this method. It is not always a practical model to generate candidate biomarkers for either diagnostic or therapy approaches.

Likewise, animal models of respiratory infection are potentially useful but have limitations also. Certain small animal strains may have limited genetic variety, such as inbred mouse lines. Thus, they might lack the diversity of host-immune responses that would mimic a human model where genomes vary widely. However, these animal models can play a crucial role in identifying molecular mechanisms of breath biomarker generation, and clearly, more work should take place to augment human clinical and cell culture studies.

While cell culture models present a "bottom-up" approach to generate candidate biomarker libraries for breath analysis, there is still room to continue to refine these systems. For instance, moving to more realistic cell culture architectures may more closely mimic tissue and organ systems as they are found *in vivo* in the human body. An example of this are the cases mentioned before where we have moved from suspension to adherent cell types, and also moving to primary cell cultures instead of immortalized cell lines. But there is further room to advance this field, and next steps are likely to include multiple co-cultures of cell types into systems that more realistically model human tissues (e.g., immune cells and epithelial cell types of the lung). Another example could include 3D cell culture systems that are gaining traction as a model system for cellular biomechanics studies.

As the field continues to evolve, it is clear that human clinical trials, animal trials, and cell culture models will produce candidate biomarkers from both a "top-down" and "bottom-up" approach. It will be critical to begin linking these candidate markers together so that we can more clearly elucidate the origin of the breath biomarkers in the clinic. This is one of the major accomplishments that will help to speed the introduction of clinic breath testing into more wide-spread practice.

ACKNOWLEDGMENTS

Partial support was provided by: Office of Naval Research (ONR) grant #N-00014-13-1-0580, The Hartwell Foundation, the NIH National Center for Advancing Translational Sciences (NCATS) through grant #UL1 TR000002, NIH award 1P30ES023513-01A1, NIH award U01 EB0220003-01 and the National Heart Lung and Blood Institute #1K23127185-01A1. The contents of this manuscript are solely the responsibility of the authors and do not necessarily represent the official views of the funding agencies.

KEYWORDS

- **bacterial**
- **breath**
- **fungus**
- **metabolomics**
- **viral**
- **VOCs**

REFERENCES

Aksenov, A. A., Gojova, A., Zhao, W., Morgan, J. T., Sankaran, S., Sandrock, C. E., & Davis, C. E., (2012). Characterization of volatile organic compounds in human leukocyte antigen heterologous expression systems: A cell's "chemical odor fingerprint." *ChemBioChem, 13*(7), 1053–1059.

Aksenov, A. A., Sandrock, C. E., Zhao, W., Sankaran, S., Schivo, M., Harper, R., Cardona, C. J., Xing, Z., & Davis, C. E., (2014a). Cellular scent of influenza virus infection. *Chembiochem., 15*(7), 1040–1048.

Aksenov, A. A., Schivo, M., Bardaweel, H., Zrodnikov, Y., Kwan, A. M., Zamuruyev, K., Cheung, W. H. K., Peirano, D. J., & Davis, C. E., (2013). Volatile organic compounds in human breath: Biogenic origin and point-of-care analysis approaches. In: *Volatile Biomarkers: Non-Invasive Diagnosis in Physiology and Medicine*, Amann, A., & Smith, D., eds., Elsevier: Boston, pp. 129–154.

Aksenov, A. A., Yeates, L., Pasamontes, A., Siebe, C., Zrodnikov, Y., Simmons, J., McCartney, M. M., Deplanque, J. P., Wells, R. S., & Davis, C. E., (2014b). Metabolite content profiling of bottlenose dolphin exhaled breath. *Anal. Chem.*, *86*(21), 10616–10624.

Alonso, A., Marsal, S., & Julià, A., (2015). Analytical methods in untargeted metabolomics: State of the art in 2015. *Frontiers in Bioengineering and Biotechnology*, *3*, 23.

Amann, A., Costello Bde, L., Miekisch, W., Schubert, J., Buszewski, B., Pleil, J., Ratcliffe, N., & Risby, T., (2014). The human volatilome: Volatile organic compounds (VOCs) in exhaled breath, skin emanations, urine, feces and saliva. *J. Breath Res.*, *8*(3), 034001.

Bansal, S., Siddarth, M., Chawla, D., Banerjee, B. D., Madhu, S. V., & Tripathi, A. K., (2012). Advanced glycation end products enhance reactive oxygen and nitrogen species generation in neutrophils in vitro. *Mol. Cell. Biochem.*, *361*(1–2), 289–296.

Bhaumik, P., Koski, M. K., Glumoff, T., Hiltunen, J. K., & Wierenga, R. K., (2005). Structural biology of the thioester-dependent degradation and synthesis of fatty acids. *Curr. Opin. Struct. Biol.*, *15*(6), 621–628.

Bos, L. D., Meinardi, S., Blake, D., & Whiteson, K., (2016). Bacteria in the airways of patients with cystic fibrosis are genetically capable of producing VOCs in breath. *J. Breath Res.*, *10*(4), 047103.

Bumatay, A., Chan, R., Lauher, K., Kwan, A. M., Stoltz, T., Delplanque, J. P., Kenyon, N. J., & Davis, C. E., (2012). Coupled mobile phone platform with peak flow meter enables real-time lung function assessment. *IEEE Sens. J.*, *12*(3), 685–691.

Chambers, S. T., Syhre, M., Murdoch, D. R., McCartin, F., & Epton, M. J., (2009). Detection of 2-pentylfuran in the breath of patients with *aspergillus fumigatus*. *Med. Mycol.*, *47*(5), 468–476.

Chen, J., Tang, J., Shi, H., Tang, C., & Zhang, R., (2016). Characteristics of volatile organic compounds produced from five pathogenic bacteria by headspace-solid phase micro-extraction/gas chromatography-mass spectrometry. *J. Basic Microbiol.*, *57*(3), 228–237.

Chitarra, G. S., Abee, T., Rombouts, F. M., Posthumus, M. A., & Dijksterhuis, J., (2004). Germination of *penicillium paneum* conidia is regulated by 1-octen-3-ol, a volatile self-inhibitor. *Appl. Environ. Microb.*, *70*(5), 2823–2829.

Costa, R., Fanali, C., Pennazza, G., Tedone, L., Dugo, L., Santonico, M., Sciarrone, D., Cacciola, F., Cucchiarini, L., Dacha, M., & Mondello, L., (2015). Screening of volatile compounds composition of white truffle during storage by GCxGC-(FID/MS) and gas sensor array analyses. *LWT-Food Sci. Technol.*, *60*(2), 905–913.

Cumeras, R., Figueras, E., Davis, C. E., Baumbach, J. I., & Gracia, I., (2015a). Review on ion mobility spectrometry. Part 1: Current instrumentation. *Analyst*, *140*(5), 1376–1390.

Cumeras, R., Figueras, E., Davis, C. E., Baumbach, J. I., & Gracia, I., (2015b). Review on ion mobility spectrometry. Part 2: Hyphenated methods and effects of experimental parameters. *Analyst*, *140*(5), 1391–1410.

Davis, C. E., Bogan, M. J., Sankaran, S., Molina, M. A., Loyola, B. R., Zhao, W., Benner, W. H., Schivo, M., Farquar, G. R., Kenyon, N. J., & Frank, M., (2010a). Analysis of volatile and non-volatile biomarkers in human breath using differential mobility spectrometry (DMS). *IEEE Sens. J., 10*(1), 114–122.

Davis, C. E., Frank, M., Mizaikoff, B., & Oser, H., (2010b). The future of sensors and instrumentation for human breath analysis. *IEEE Sens. J., 10*(1), 3–6.

Dryahina, K., Sovova, K., Nemec, A., & Spanel, P., (2016). Differentiation of pulmonary bacterial pathogens in cystic fibrosis by volatile metabolites emitted by their in vitro cultures: Pseudomonas aeruginosa, *staphylococcus aureus, stenotrophomonas maltophilia* and the *burkholderia cepacia* complex. *J. Breath Res., 10*(3), 037102.

Filipiak, W., Mochalski, P., Filipiak, A., Ager, C., Cumeras, R., Davis, C. E., Agapiou, A., Unterkofler, K., & Troppmair, J., (2016). A compendium of volatile organic compounds (VOCs) released by human cell lines. *Curr. Med. Chem., 23*(20), 2112–2131.

Filipiak, W., Sponring, A., Filipiak, A., Ager, C., Schubert, J., Miekisch, W., Amann, A., & Troppmair, J., (2010). TD-GC-MS analysis of volatile metabolites of human lung cancer and normal cells *in vitro. Cancer Epidem. Biomar., 19*(1), 182–195.

Haick, H., Broza, Y. Y., Mochalski, P., Ruzsanyi, V., & Amann, A., (2014). Assessment, origin, and implementation of breath volatile cancer markers. *Chem. Soc. Rev., 43*(5), 1423–1449.

Hendriks, M. M. W. B., Eeuwijk, F. A. V., Jellema, R. H., Westerhuis, J. A., Reijmers, T. H., Hoefsloot, H. C. J., & Smilde, A. K., (2011). Data-processing strategies for metabolomics studies. *Trends Analyt. Chem., 30*(10), 1685–1698.

Humphris, S. N., Bruce, A., Buultjens, E., & Wheatley, R. E., (2002). The effects of volatile microbial secondary metabolites on protein synthesis in *serpula lacrymans. FEMS Microbiol. Lett., 210*(2), 215–219.

Junger, M., Vautz, W., Kuhns, M., Hofmann, L., Ulbricht, S., Baumbach, J. I., Quintel, M., & Perl, T., (2012). Ion mobility spectrometry for microbial volatile organic compounds: a new identification tool for human pathogenic bacteria. *Appl. Microbiol. Biotechnol., 93*(6), 2603–2614.

Kramer, R., & Abraham, W. R., (2012). Volatile sesquiterpenes from fungi: What are they good for? *Phytochem. Rev., 11*(1), 15–37.

Kwan, A. M., Fung, A. G., Jansen, P. A., Schivo, M., Kenyon, N. J., Delplanque, J. P., & Davis, C. E., (2015). Personal lung function monitoring devices for asthma patients. *IEEE Sens. J., 15*(4), 2238–2247.

Lemfack, M. C., Nickel, J., Dunkel, M., Preissner, R., & Piechulla, B., (2014). mVOC: a database of microbial volatiles. *Nucleic Acids Res., 42*(D1), D744–D748.

Li, N. X., Alfiky, A., Vaughan, M. M., & Kang, S., (2016). Stop and smell the fungi: Fungal volatile metabolites are overlooked signals involved in fungal interaction with plants. *Fungal Biol. Rev., 30*(3), 134–144.

Mansoor, J. K., Schelegle, E. S., Davis, C. E., Walby, W. F., Zhao, W. X., Aksenov, A. A., Pasamontes, A., Figueroa, J., & Allen, R., (2014). Analysis of volatile compounds in exhaled breath condensate in patients with severe pulmonary arterial hypertension. *PLOS One, 9*(4), e95331.

Matysik, S., Herbarth, O., & Mueller, A., (2008). Determination of volatile metabolites originating from mold growth on wallpaper and synthetic media. *J. Microbiol. Meth., 75*(2), 182–187.

Michael, S., Alexander, A. A., Hamzeh, B., Weixiang, Z., Nicholas, J. K., & Cristina, E. D., (2011). Building biomarker libraries with novel chemical sensors: correlating differential mobility spectrometer signal outputs with mass spectrometry data. *IOP Conf. Ser. Mater. Sci. Eng.*, *18*(21), 212003.

Mogayzel, P. J., Jr., Naureckas, E. T., Robinson, K. A., Mueller, G., Hadjiliadis, D., Hoag, J. B., Lubsch, L., Hazle, L., Sabadosa, K., & Marshall, B., (2013). Pulmonary clinical practice guidelines, C., cystic fibrosis pulmonary guidelines. Chronic medications for maintenance of lung health. *Am. J. Respir. Crit. Care Med.*, *187*(7), 680–689.

Nagana Gowda, G. A., Zhang, S., Gu, H., Asiago, V., Shanaiah, N., & Raftery, D., (2008). Metabolomics-based methods for early disease diagnostics: A review. *Expert Rev. Mol. Diagn.*, *8*(5), 617–633.

Neerincx, A. H., Geurts, B. P., Habets, M. F., Booij, J. A., Van Loon, J., Jansen, J. J., Buydens, L. M., Van Ingen, J., Mouton, J. W., Harren, F. J., Wevers, R. A., Merkus, P. J., Cristescu, S. M., & Kluijtmans, L. A., (2016). Identification of *pseudomonas aeruginosa* and *aspergillus fumigatus* mono-and co-cultures based on volatile biomarker combinations. *J. Breath Res.*, *10*(1), 016002.

Nemcovic, M., Jakubikova, L., Viden, I., & Farkas, V., (2008). Induction of conidiation by endogenous volatile compounds in Trichoderma spp. *FEMS Microbiol. Lett.*, *284*(2), 231–236.

Nording, M. L., Yang, J., Hegedus, C. M., Bhushan, A., Kenyon, N. J., Davis, C. E., & Hammock, B. D., (2010). Endogenous levels of five fatty acid metabolites in exhaled breath condensate to monitor asthma by high-performance liquid chromatography: Electrospray tandem mass spectrometry. *IEEE Sens. J.*, *10*(1), 123–130.

Peirano, D., Aksenov, A., Pasamontes, A., & Davis, C., (2013). Approaches for establishing methodologies in metabolomic studies for clinical diagnostics. In: *Medical Applications of Artificial Intelligence*, CRC Press, pp. 279–306.

Peled, N., Barash, O., Tisch, U., Ionescu, R., Broza, Y. Y., Ilouze, M., Mattei, J., Bunn, P. A. Jr., Hirsch, F. R., & Haick, H., (2013). Volatile fingerprints of cancer specific genetic mutations. *Nanomedicine*, *9*(6), 758–66.

Peralbo-Molina, A., Calderón-Santiago, M., Priego-Capote, F., Jurado-Gámez, B., & Castro, M. D., (2016). Metabolomics analysis of exhaled breath condensate for discrimination between lung cancer patients and risk factor individuals. *J. Breath Res.*, *10*(1), 016011.

Peralbo-Molina, A., Calderón-Santiago, M., Priego-Capote, F., Jurado-Gámez, B., & Luque de Castro, M. D., (2015). Development of a method for metabolomic analysis of human exhaled breath condensate by gas chromatography–mass spectrometry in high resolution mode. *Anal. Chim. Acta*, *887*, 118–126.

Robacker, D. C., & Lauzon, C. R., (2002). Purine metabolizing capability of *enterobacter agglomerans* affects volatiles production and attractiveness to Mexican fruit fly. *J. Chem. Ecol.*, *28*(8), 1549–63.

Schivo, M., Aksenov, A. A., Linderholm, A. L., McCartney, M. M., Simmons, J., Harper, R. W., & Davis, C. E., (2014). Volatile emanations from in vitro airway cells infected with human rhinovirus. *J. Breath Res.*, *8*(3), 037110.

Schulz, S., & Dickschat, J. S., (2007). Bacterial volatiles: the smell of small organisms. *Nat. Prod. Rep.*, *24*(4), 814–842.

Shnayderman, M., Mansfield, B., Yip, P., Clark, H. A., Krebs, M. D., Cohen, S. J., Zeskind, J. E., Ryan, E. T., Dorkin, H. L., Callahan, M. V., Stair, T. O., Gelfand, J. A., Gill, C. J., Hitt, B., & Davis, C. E., (2005). Species-specific bacteria identification using differential mobility spectrometry and bioinformatics pattern recognition. *Anal. Chem.*, *77*(18), 5930–5937.

Smith, D., Sovova, K., Dryahina, K., Dousova, T., Drevinek, P., & Spanel, P., (2016). Breath concentration of acetic acid vapour is elevated in patients with cystic fibrosis. *J. Breath Res.*, *10*(2), 021002.

Sobus, J. D. P., (2013). Mathematical and statistical approaches for interpreting biomarker compounds in exhaled breath. In: *Volatile Biomarkers: Non-invasive Diagnosis in Physiology and Medicine*, Amann, A., & Smith, D., eds. Elsevier, Boston, pp. 3–18.

Sofia, M., Maniscalco, M., De Laurentiis, G., Paris, D., Melck, D., & Motta, A., (2011). Exploring airway diseases by NMR-based metabonomics: A review of application to exhaled breath condensate. *J. Biomed. Biotechnol*, pp. 7, http://dx.doi.org/10.1155/2011/403260.

Strand, N., Bhushan, A., Schivo, M., Kenyon, N. J., & Davis, C. E., (2010). Chemically polymerized polypyrrole for on-chip concentration of volatile breath metabolites. *Sensor. Actuat. B-Chem.*, *143*(2), 516–523.

Tait, E., Perry, J. D., Stanforth, S. P., & Dean, J. R., (2014a). Bacteria detection based on the evolution of enzyme-generated volatile organic compounds: determination of *listeria monocytogenes* in milk samples. *Anal. Chim. Acta*, *848*, 80–87.

Tait, E., Perry, J. D., Stanforth, S. P., & Dean, J. R., (2014b). Identification of volatile organic compounds produced by bacteria using HS-SPME-GC-MS. *J. Chromatogr. Sci.*, *52*(4), 363–373.

Troy, N. M., & Bosco, A., (2016). Respiratory viral infections and host responses, insights from genomics. *Resp. Res.*, *17*, p. 156.

Vahdatzadeh, M., Deveau, A., & Splivallo, R., (2015). The role of the microbiome of truffles in aroma formation: a meta-analysis approach. *Appl. Environ. Microb.*, *81*(20), 6946–6952.

Van der Greef, J., & Smilde, A. K., (2005). Symbiosis of chemometrics and metabolomics: past, present, and future. *J. Chemometrics*, *19*(5–7), 376–386.

Van der Schee, M. P., Hashimoto, S., Schuurman, A. C., Repelaer Van Driel, J. S., Adriaens, N., Van Amelsfoort, R. M., Snoeren, T., Regenboog, M., Sprikkelman, A. B., Haarman, E. G., Van Aalderen, W. M. C., & Sterk, P. J., (2014). Altered exhaled biomarker profiles in children during and after rhinovirus-induced wheeze. *Eur. Respir. J.,* 440–448.

Vanhoutte, P. M., (2013). Airway epithelium-derived relaxing factor: myth, reality, or naivety? *Am. J. Physiol. Cell Physiol.*, *304*(9), 813–820.

Yamada, Y., Kuzuyama, T., Komatsu, M., Shin-Ya, K., Omura, S., Cane, D. E., & Ikeda, H., (2015). Terpene synthases are widely distributed in bacteria. *Proc. Natl. Acad. Sci. USA*, *112*(3), 857–862.

Yi, L., Dong, N., Yun, Y., Deng, B., Ren, D., Liu, S., & Liang, Y., (2016). Chemometric methods in data processing of mass spectrometry-based metabolomics: A review. *Anal. Chim. Acta*, *914*, 17–34.

Zrodnikov, Y., Zamuruyev, K., Pedersen, J. D., Fung, A. G., Peirano, D. J., Schirle, M. J., Panigrahy, A., Pasamontes, A., Cheung, W. H. K., Aksenov, A. A., Schivo, M., Kenyon,

N. J., Delplanque, J. P., & Davis, C. E., (2013). Design criteria for portable point-of-care breath analysis systems. 2013 Transducers & Eurosensors XXVII: 17th International Conference on Solid-State Sensors, *Actuators and Microsystems*, 1629–1632.

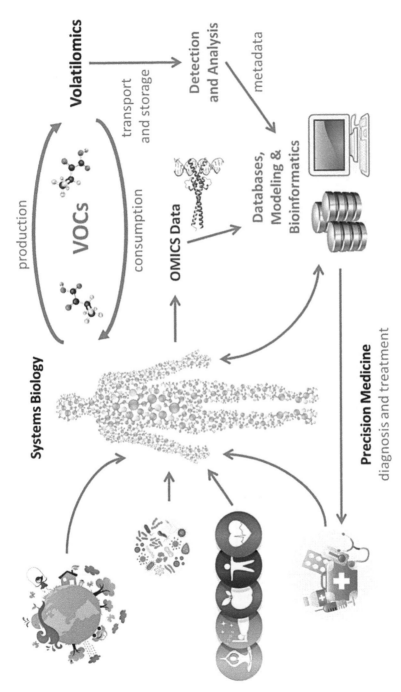

FIGURE 1.1 Systems biology, volatilomics and precision medicine relations.

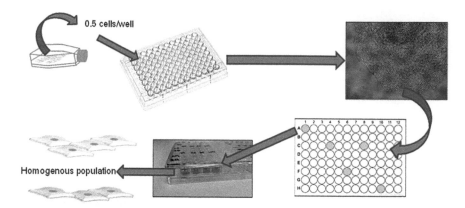

FIGURE 3.1   Schematic protocol to isolate cell clones from cancer cell lines.

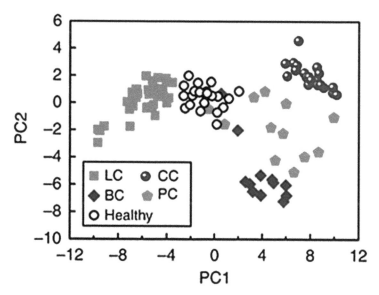

FIGURE 4.2   PCA scores plot showing the discrimination between lung cancer (LC), colorectal cancer (CC), breast cancer (BC), prostate cancer (PC), and healthy controls (adapted from Peng, et al. (2010). Br. J. Cancer 103(4), 542–551. With permission).

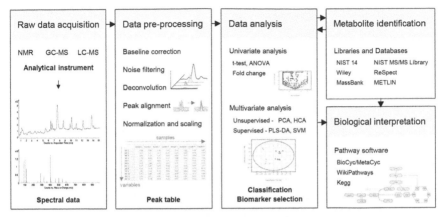

FIGURE 5.1   Typical steps needed for breath biomarker discovery.

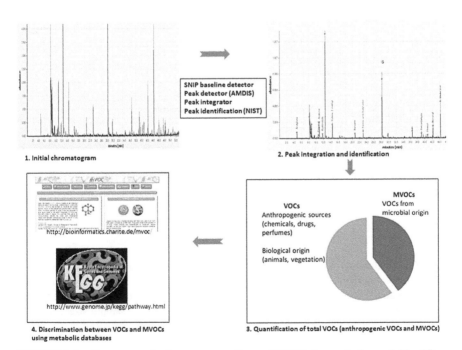

FIGURE 6.5   Treatment of a chromatogram from a WWTP air sample for peak identification and separation of MVOCs from VOCs with online databases.

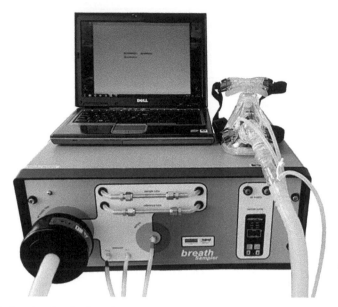

FIGURE 7.3   An alveolar breath sampling system based on a modified CPAP device.

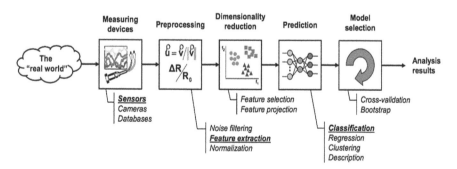

FIGURE 8.1   General workflow of a bioinformatic solution or pattern recognition problem. First, measuring devices are needed (e.g., NMR, GC-MS, LC-MS) and the obtained data is pre-processed. The number of features extracted from the data (data dimensionality) is often reduced due to its large value with respect to the often small sample size. Then, a qualitative or quantitative prediction model is built with a set of known samples (training set) and evaluated (validation set) to both select the optimum model and estimate the performance (From Nagle et al., (1998). *IEEE Spectrum., 35*(9), 22–31. With permission.).

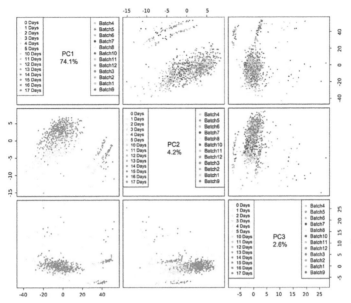

FIGURE 8.3 PCA scoreplot of urine samples analysed by LC-MS organized either by batch or day of measurement. The numbers in the diagonal plot represent the variance captured by each principal component (From Fernández-Albert et al., (2014). *Bioinformatics., 30,* 2899–2905. With permission.).

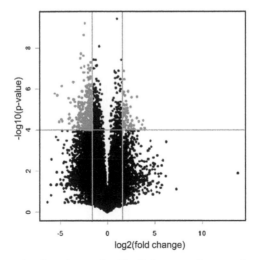

FIGURE 8.6 Example of a volcano plot. The lighter samples are selected, since the value of their p-value is lower than a stated threshold and their ES (in the form of a fold change) is located outside the region defined by another threshold, specifically ±ES threshold (From Patti et al., (2012). Nat. Rev. Mol. Cell. Biol., 13, 263–269. With permission.).

FIGURE 9.1   The main window of the KNApSAcK Family Databases (http://kanaya.naist.jp/KNApSAcK_Family/).

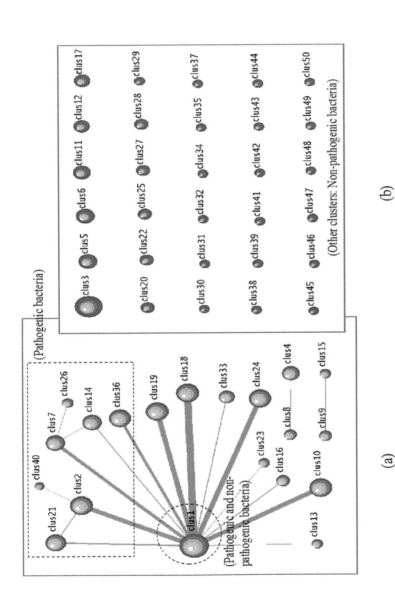

FIGURE 9.7 Hierarchical graph of DPClus clustering result. (a) Connected nodes – nodes enclosed by dotted rectangle are consisting of only pathogenic bacteria, the only node enclosed by the dotted circle is consisting of both pathogenic and non-pathogenic bacteria and the rest nodes are consisting of only non-pathogenic bacteria. (b) Independent nodes – all nodes are consisting of non-pathogenic bacteria (Adapted from Abdullah, et al. (2015). Biomed Res. Int., 139–254.).

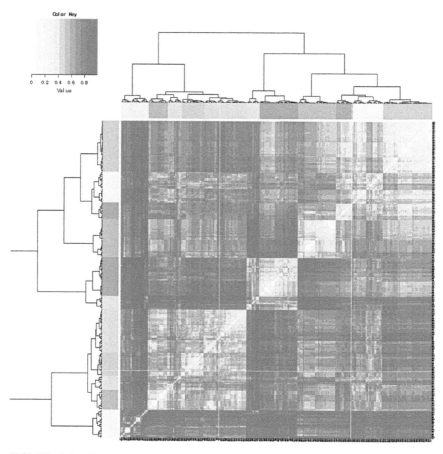

FIGURE 9.9 Heatmap clustering of VOCs based on chemical structure similarity determined by Tanimoto coefficient (From Abdullah, et al. (2015). Biomed Res. Int., 139–254.).

CHAPTER 6

CHARACTERIZING OUTDOOR AIR USING MICROBIAL VOLATILE ORGANIC COMPOUNDS (MVOCs)

SONIA GARCIA-ALCEGA and FRÉDÉRIC COULON

Cranfield University, School of Water, Energy and Environment, Cranfield, MK43 0AL, UK, Tel.: +44 (0) 1234 75 4981, E-mail: f.coulon@cranfield.ac.uk

ABSTRACT

Exposure to bioaerosols containing airborne microorganisms and their by-products from outdoor environments such as industrial, urban or agricultural sites is of great concern as it is linked to adverse health effects in humans including respiratory diseases and infections. The risk exposure from outdoor emissions is difficult to quantify in real-time as the microbial concentration in air is low and varies depending on meteorological factors, anthropogenic activities, and sampling conditions. In addition, the collection of sufficient amount of sample to generate statistically distinguishable and reproducible patterns to characterize and quantify bioaerosols is still a challenge, and this analysis cannot be performed in real time yet. Microbial volatile organic compounds (MVOCs) can be used to chemically characterize ambient bioaerosols and identify pathogens early in air overcoming the inherent limitations of culturing. This book chapter aims to critically review the sampling techniques and analytical approaches that are currently available for the study of MVOCs from industrial, agricultural and rural emissions. Current challenges in MVOCs sample collection, analytical and speciation analysis are addressed, and recommendation for the implementation of a rapid, reproducible and sensitive analytical framework for fingerprinting bioaerosols is provided.

6.1   INTRODUCTION

Bioaerosols are defined as the biological particles present in aerosols that originate from plants, animals and microbes (Gómez-Domenech et al., 2010). They are ubiquitous in the environment, and their small particle size ranging between 10 nm and 2.5 µm means that they easily propagate in air (Prospero et al., 2005). Among bioaerosols, some are pathogens, and recent studies have shown that chronic exposure can induce adverse human health effects such as respiratory diseases, infections (Sharma et al., 2015) and asthma (Ghosh et al., 2015; O'Connor et al., 2015). Consequently, bioaerosols emissions from industrial, urban and agricultural environments such as wastewater treatment plants (WWTP), composting facilities and other bio-waste processing plants have raised public concerns as the concentrations can be potentially high and cause a negative impact on local air quality (Macklin et al., 2011; Pankhurst et al., 2012).

Current bioaerosol monitoring methods can be classified into cultural, molecular and chemical techniques (Kim et al., 2009, 2016) (Table 6.1). Although less than 1% of viable microbes are cultivable under standard laboratory conditions, conventional culturing methods have been widely used for bioaerosol studies. These techniques are easy to use and can allow microbial identification at species level, but the results often underestimate the microbial quantity and diversity, are labor intensive and not very reproducible (Oliver, 2005). Molecular techniques are more expensive but are very reproducible allowing the microbial detection at very low concentrations. These techniques include immunochemistry, flow cytometry, microscopy (Mandal and Brandl, 2011), pyrosequencing, DNA and RNA techniques (Yoo et al., 2016) as well as phospholipid fatty acids (PLFAs) (Pankhurst et al., 2012; Willers et al., 2015). Molecular techniques have the advantage compared to culture techniques of being reproducible and that microbial species can be identification at species level (except with PLFAs analysis that are only specific to a certain microbial group and they give information about the structure of the microbial community). The disadvantages of molecular techniques are, for instance, that the extraction protocols are long and expensive and the data cannot be obtained in real time. Chemical techniques are getting increased interest for the characterization of bioaerosols as they offer reproducible, accurate and reliable analysis, are sensitive allowing the detection of the chemicals in low concentrations and are rapid, inexpensive and can perform real time analysis. Chemical markers used for bioaerosol characterization include biological particles, species-specific proteins, and MVOCs. Chemical techniques focusing on the discrimination between biological and non-biological particles are the fluorescent aerodynamic particle sizer (FLAPS) (Pankhurst et al., 2012), Spectral Intensity Bioaerosol Sensor (SIBS) (Schmidt and Bauer, 2010) and Wideband Integrated Bioaerosol Sensor (WIBS) (Toprak and Schnaiter, 2013). These sensors operate using light scattering and fluorescence detection and can work over long periods of time offering real time results but disadvantageously are very heavy and produce a large amount of data, which is difficult to process. Carbon nanotube based biosensor is another sensing technique that monitors continuously two different airborne fungal species simultaneously (Kim et al., 2016). This device is not designed to be portable yet, and it needs to be configured to detect more microbial species. There are also biosensors based on microfluidic techniques that collect the air sample directly into the liquid of a microfluidic cartridge and concentrate it (from L to µL) prior the detection with an integrated biosensing (Pardon et al., 2015; Sandström et al., 2015). Mass spectrometers are also applied for the chemical characterization of bioaerosols. The Bioaerosol Mass Spectrometer (BMAS) discriminates between microbial species by the mass spectrum analysis of species-specific proteins and peptides (Srivastava et al., 2005). Conventional GC-MS allow the identification and quantification of VOCs and MVOCs by their mass spectra (Lavine et al., 2012). Raman spectroscopy techniques are used to obtain information about the average size of the molecules although they do not give information about the sample composition. Fourier Transform infrared spectroscopy (FT-IR) and near-infrared spectroscopy (NIR) have also been used in bioaerosol studies to differentiate the molecules by the absorption of infrared light (Quintelas et al., 2015; Siddiquee et al., 2015). Although no single method is preferred over another, the selection is generally determined by the objectives of the research, and the difference in sampling methodologies can hamper data comparison.

MVOCs are secondary metabolites resulting from microbial fermentation. They are characterized by low boiling points, high vapor pressures and low molecular weights (Schenkel et al., 2015). As microbial communities exhibit different MVOCs depending on which environment they are in (Konuma et al., 2015) and fingerprinting analytical techniques have evolved considerably, chemical analysis of MVOCs has the potential to be a reliable and rapid approach for the characterization of ambient bioaerosols (Lemfack et al., 2014; Mason et al., 2010). Furthermore, a rapid detection of pathogenic fungi and bacteria present in the air could be implemented by the detection of selective MVOCs (Araki et al., 2012).

MVOCs are produced by all the microorganisms, and not all compounds are uniquely produced by a single microbial species. Species-specific MVOCs have a potential to be used as chemical biomarkers for the selective detection and identification of bacterial and fungal species in ambient air, being untargeted metabolomics, therefore, a potential way forward for the identification of new compounds. Although human risk exposure to MVOCs is linked to the type and concentration of the microbial species, to date is still difficult to perform a real time identification and quantification of the microorganisms present in air (Albrecht et al., 2008; Kuske et al., 2005; Müller et al., 2004).

This book chapter aims to critically review the sampling techniques and analytical approaches that are available at the moment for the study of microbial VOCs from industrial, agricultural and rural emissions. Current challenges on MVOCs sample capture, analytical and speciation analysis are addressed, and directions for the implementation of a rapid, reproducible and sensitive working mechanism for fingerprinting bioaerosols looking at MVOCs are given.

6.2 MICROBIAL VOLATILE ORGANIC COMPOUNDS

6.2.1 MVOCs IN THE ENVIRONMENT

The analysis of MVOCs has already been applied in different fields such as in forensics and security to detect drugs, explosives or warfare (Eiceman and Stone, 2004), for fungal detection at composting sites (Fisher et al., 2004), as well as in health care to detect *Aspergillus fumigatus* in breath samples (Chambers et al., 2011) or to diagnose Crohn's disease from urine (Cauchi et al., 2015). However, to date, most of the studies have been focused on indoor built environments, and there is limited characterization of MVOCs profiles from outdoor environments apart from animal farms or bio-waste facilities. There is a need therefore of baseline fingerprint of different environments to understand the true contribution of anthropogenic activities and potential health issues as well as seasonal changes of

TABLE 6.1 Advantages and Disadvantages of Available Sampling Techniques to Study Bioaerosols

Techniques	Target compounds/molecules	Advantages	Disadvantages	References
Cultural techniques				
- Plating - Microscopy - Flow cytometry	- CFU - Number of live/dead cells	- Easy to use - Identification at species level	- Labor intensive - Expensive - Low reproducibility - No real time - Underestimation	(Kim et al., 2009, 2016; Oliver, 2005; Usachev et al., 2012)
Molecular techniques				
- Immunochemistry - PCR, Q-PCR - Pyrosequencing - DNA/RNA-based techniques - Phospholipids analysis	- Proteins - DNA - PLFAs	- Identification at species level - Reproducible analysis - Detection at low level	- Expensive - Long laboratory protocols - No real time	(Pankhurst et al., 2012; Srivastava et al., 2005; Willers et al., 2015; Yoo et al., 2016)
Chemical techniques				
- FLAPS - Sensing techniques - MVOCs - FT-IR - FT-NIRS - Raman spectroscopy - Mass spectrometry	-Particles -MVOCs - Proteins	- Reproducible analysis - Detection at low level - Accurate and reliable - Rapid, non-destructive, and inexpensive - Possibility of real time	- Biosensors not easy to transport - MVOCs collection techniques are easily portable	(Kim et al., 2016; Pardon et al., 2015; Schmidt and Bauer, 2010; Srivastava et al., 2005; Toprak and Schnaiter, 2013)

Table modified from Garcia-Alcega et al. (2016).

MVOCs and differences in emissions depending on the source or dynamics of the site activity.

MVOCs are present in outdoor air at trace levels, and as a result of their high diversity, they are grouped in 13 different chemical groups (Table 6.2) (Schenkel et al., 2015): alkanes, alkenes, nitrogen compounds, sulphur compounds, aldehydes, ketones, alcohols, organic acids, ethers, esters, furans, aromatic compounds, and terpenes. MVOCs differ in their physicochemical properties as well as in their chemical structure, so efficient sampling methods are required to be able to collect as many compounds as possible.

Table 6.3 summarizes the most common MVOCs identified from outdoor (compost facilities, municipal solid waste management, and a WWTP) and indoor (houses, buildings and broiler sheds) ambient air (García-Alcega et al., 2016). In every study, samples were collected with different techniques i.e. Tenax TA, GR tubes, Carbopack B with sampling times varying between 30 min to 3–5 h and flow rates between 55–100 mL min^{-1} and in one of the studies the sampling technique was not mentioned at all. The wide range of MVOCs found between studies can be attributed to the inconsistency in sample collection devices and sampling settings (flow rate and sampling time). Monitoring and measuring MVOCs in an indoor environment is easier than outdoors as the concentrations are usually higher and it is a more constant environment. Outdoor MVOCs concentrations instead are much lower, and there are several co-founding factors and environmental parameters that influence on the sample capture such as weather conditions or activity that is being carried out on the site. Due to the low outdoor environment concentrations, it is difficult to differentiate MVOCs profiles among sites or to identify a contaminated site from a non-contaminated site.

Apart from this, researchers do not analyze and report same MVOCs (García-Alcega et al., 2016), and MVOC contaminant concentration threshold and concentration limits are not consistent. For example, in a study of indoor air from buildings, Lorenz et al. (2002) identified the MVOCs 1-octen-3-ol, dimethyl disulfide and 3-methylfuran as the main indicators of microbial growth. The authors determined that there is an indoor microbial source of contamination when the detection of one of these MVOCs is present at concentrations above 50 ng m^{-3} or when the sum of eight MVOCs (1-octen-3-ol, 3-methylfurane, dimethyl disulfide, 3-methyl-1-butanol, 2-pentanol, 2-hexanone, 2-heptanone, 3-octanone)

TABLE 6.2 Chemical Group, Physicochemical Properties and Microbial Origin of the Predominant MVOCs Present in the Environment

Chemical group	Compound	Chemical formula	Molecular weight (g mol⁻¹)	log K_{ow}	Boiling point (°C) at 101.3 kPa	Vapor pressure (kPa at 25°C)	Microbial origin
Acids	Butanoic acid	$C_4H_8O_2$	88.1	0.8	163.75	0.74	Bacteria and fungi
	Propanoic acid	$C_3H_6O_2$	74.07	0.3	141.15	0.47	Bacteria
Aldehydes	Acetaldehyde	C_2H_4O	44.05	-0.3	20.2	101	Bacteria and fungi
	Furfural	$C_5H_4O_2$	96.08	0.41	162	0.15	Bacteria and fungi
Ethers	2-methylfuran	C_5H_{60}	82.1	1.85	65	23.48	Bacteria and fungi
	3-Methylfuran	C_5H_6O	82.1	1.91	65-66	21.46	Bacteria and fungi
Sulphur and nitrogen compounds	Dimethyl-sulfide	C_2H_6S	62.134	0.977	188.8	53.7	Bacteria and fungi
	Dimethyl disulfide	$C_2H_6S_2$	94.19	1.77	109.8	3.83	Bacteria and fungi
	2-isopropyl-3-methoxypyrazine	$C_8H_{12}N_2O$	152.2	2.37	210.8±30.0	0.036	Bacteria
Ketones	2-heptanone	$C_7H_{14}O$	114.19	2.03	150.6-151.5	0.213-0.28	Bacteria and fungi
	2-hexanone	$C_6H_{12}O$	100.16	1.38	126-128	1.47,0.36	Bacteria and fungi
	3-Octanone	$C_8H_{16}O$	128.21	2.22	157-162	0.267	Bacteria and fungi
Terpenes	Geosmin	$C_{12}H_{22}O$	182.31	3.57	252.4±8.0	0.00041	Bacteria
	Borneol	$C_{10}H_{18}O$	154.25	2.3	213	0.009	Bacteria
	2-methylisorbenol	$C_{11}H_{20}O$	168.28	3.31	208.7±8.0	0.0065	Bacteria
	β-Caryophyllene	$C_{15}H_{24}$	204.35	NA	254	NA	Bacteria
	α-Pinene	$C_{10}H_{16}$	136.23	2.8	155	0.4	Bacteria and fungi
	Camphene	$C_{10}H_{16}$	136.23	3.3	159	NA	Fungi
	Camphor	$C_{10}H_{16}O$	152.23	2.2	209	0.53	Bacteria and fungi

TABLE 6.2 (Continued)

Chemical group	Compound	Chemical formula	Molecular weight (g mol⁻¹)	log K_{ow}	Boiling point (°C) at 101.3 kPa	Vapor pressure (kPa at 25°C)	Microbial origin
Alcohols	2-methyl-1-propanol	$C_4H_{10}O$	74.12	0.65–0.83	108	1.33	Bacteria and fungi
	2-methyl-1-butanol	$C_5H_{12}O$	88.15	1.29	128	0.416	Bacteria and fungi
	3-methyl-1-butanol	$C_5H_{12}O$	88.15	1.16	130.5	0.316	Bacteria and fungi
	3-methyl-2-butanol	$C_5H_{12}O$	88.15	1.28	111.5	1.22	Bacteria
	3-octanol	$C_8H_{18}O$	130.23	2.73	169	0.068	Fungi
	1-octen-3-ol	$C_8H_{16}O$	128.21	2.6	180	0.071	Bacteria and fungi
	2-octen-1-ol	$C_8H_{16}O$	128.21	2.59	195.8±8.0	0.014	Bacteria and fungi
	2-pentanol	$C_5H_{12}O$	88.15	1.19	119.0–119.3	0.815	Bacteria and fungi
	Methanol	CH_4O	32.04	−0.5	64.7	32	Bacteria and fungi

*log K_{ow} = Octanol-Water partition coefficient.
Table modified from Garcia-Alcega et al. (2016).

TABLE 6.3 Most Frequent MVOCs in Outdoor and Indoor Environments and Range of Concentrations (ng m^{-3})

MVOCs	Outdoor environments			Indoor environments			
	Compost facilities[a]	MSW[b]	WWTP[c]	Normal buildings[d]	Living environments[d,e]	Problem buildings[f]	Broiler sheds[g]
	Tenax TA and GR tubes (sampling conditions n.s.)	Tenax GR tubes filled with adsorption resin (200 mg) @ 100 mL/min; sampling time n.s.	Tenax GR tubes filled with adsorption resin (200 mg) @ 100 mL/min; sampling time n.s.	Tenax TA tubes (30 min @ 100 mL/min) TD tubes (adsorbent and time n.s.)	300–400 mg Tenax TA-Carbopack B tubes (45–55 mL/min during 3–5 h) TD tubes 30 min @ 100 mL/min (adsorbent and time n.s.)	n.s.	Tenax TA tubes (30 min @ 100 mL/min)
				Sampling technique			
Ethanol	na	250	na	na	na	na	na
2-propanol	na	120	na	na	na	na	na
2-methyl-1-butanol	170–1400	na	na	na	na	na	na
2-methyl-1-propanol	na	na	na	340–1380	3000–10400	nd–1740	na
3-methyl-1-butanol	300–35000	na	na	8700–110000	3000	175–260000	nd–25000
3-methyl-2-butanol	nd–70	na	na	nd–160	3610	190–1190	na
3-octanol	nd–140	na	na	nd–40	5330–8800	nd–8860	na
1-octen-3-ol	nd–1900	na	na	nd–7000	5240–11800	nd–904000	300–6000
2-octen-1-ol	nd–6820	na	na	nd–14000	5240–21500	1560–266000	na
2-pentanol	na	na	na	1700	3610–4800	nd–1400	na
2-methylfuran	75–1500	na	na	na	6300	na	na
3-methylfuran	nd–110	na	na	nd–160	3360	nd–1800	na
2-penthylfuran	85–1240	na	na	na	5100	na	na
2,3-butanedione	na	90	na	na	na	na	3000–324000

TABLE 6.3 *(Continued)*

MVOCs	Outdoor environments			Indoor environments			
	Compost facilities [a]	MSW [b]	WWTP [c]	Normal buildings [d]	Living environments [d,e]	Problem buildings [f]	Broiler sheds [g]
2-butanone	na	na	13700	na	na	na	na
3-Hydroxy-2-butanone	na	140	na	na	na	na	na
2-heptanone	nd–3000	na	na	nd–1200	4670–16900	nd–97	na
2-hexanone	nd–800	na	na	nd	4100	25–8800	na
3-octanone	nd–2000	na	na	nd–3000	5240–11600	nd–3020	na
Acetic acid	na	60	na	na	na	na	na
Ethyl acetate	na	110	na	na	na	na	na
Nonane	na	80	na	na	na	na	na
Decane	na	110	na	na	na	na	na
Undecane	na	320	nd	na	na	na	na
borneol	160–7000	na	na	na	6900	na	na
geosmin	nd–10	na	na	nd–50	6000–7460	nd–550	na
2-methyli-sorbenol	nd–1180	na	na	nd–560	6880	nd–2800	na
dimethylsulfide	<50–3300	na	26400	na	1700	na	nd–1700
dimethyl disulfide	nd–6000	na	22500	nd–710	3850–263000	16–90	nd–263000
2-isopropyl-3-methoxy-pyrazine	nd–340	na	na	nd–3	6220	nd–9500	na

[a] Compost facilities (Fisher et al, 2008; Müller et al., 2004); [b] MSW = Municipal solid waste treatment (Lehtinen et al., 2013); [c] WWTP = Waste water treatment plant (at sludge dewatering site) (Lehtinen and Veijanen, 2011); [d] Normal buildings = Without damp problems or non-complaint areas (Fisher et al., 2004; Lorenz et al., 2002; Schleibinger et al., 2008); [e] Living environments = Houses (Fisher et al., 2004; Lorenz et al., 2002; Murphy et al., 2014); [f] Problem buildings = Buildings with damp problems (Korpi et al., 2009; Ström et al., 1994); [g] Broiler sheds (Murphy et al., 2014); *n.s. = not specified; na = not available; nd = not detected.

together with at least one of the 3 main MVOCs indicators of microbial growth equals or exceeds 500 ng m^{-3}. Opposite to this, Korpi et al. (2009) suggested other limits for 3-methylfurane ($\geq$ 200) and 1-octene-3-ol and 3-methyl-1-butanol $\geq$ 10000 ng m^{-3}. This author also listed another MVOCs different from the ones reported by Lorenz et al. (geosmin $\geq$ 50, 2-isopropyl-3- methoxypyrazine $\geq$ 400, 2-methyl1-propanol and 2-methylisoborneol $\geq$ 1500, 2-octen-1-ol $\geq$ 15000 ng m^{-3}).

6.2.2 SPECIES-SPECIFIC MVOCs

The identification of species-specific MVOCs is difficult because first, not all the MVOCs present in the air have solely microbial origin and second because most of the reported MVOCs with microbial origin are not species-specific (Bos et al., 2013; Dunkel, et al. 2009; Kuske et al., 2005; Schleibinger et al., 2008). The microbial origin of the most frequent outdoor MVOCs is summarized in Table 6.3. 2-hexanone and 3-methyl-1-butanol are MVOCs uniquely related to microbial sources. However, they are produced during the metabolism of every fungi and bacteria and therefore are not specific to any microbial species. 3-octanol for instance is only emitted by fungi, and other MVOCs like 3-methyl-2-butanol, geosmin, borneol, 2-methyl-isorbenol, and 2-isopropyl-3-methoxypyrazine are uniquely produced by bacteria. Nevertheless, none of them can be attributed to any specific microorganism. There are inconsistencies about which is the microbe producer of which MVOCs. As an example, 3-octanone was reported to be only emitted by *Aspergillus fumigatus* but lately, it has been discovered that *Aspergillus flavus* also produces it (Stotzky and Schenck, 1976). Additionally, Gao et al. (2002) suggested that 2-pentyl furan was a metabolite exclusively from *Aspergillus fumigatus,* and not long ago other researchers revealed that there are other fungi that also produce it (*Aspergillus flavus, Aspergillus niger, Aspergillus terreus, Scedosporium apiospermum* and *Fusarium spp* respectively) as well as the bacteria *Streptococcus pneumonia* (Chambers et al., 2011; Syhre et al., 2008).

   Vishwanath et al. (2011) tried to discriminate species specific MVOCs from anthropogenic VOCs present in dust from houses and waste management facilities. They did not succeed in their study due to the ambiguity of MVOCs and the lack of a certified reference material reporting MVOCs concentrations in environmental samples. In another recent study, Choi et al. (2016) analyzed in parallel dust and air samples from buildings and houses to investigate the anthropogenic and microbial sources of 28 VOCs that are considered to have a microbial origin. They concluded that the $\sum 28$ VOCs identified in the samples which were commonly accepted as MVOCs, were, in fact, more associated to anthropogenic sources such as the phthalates 2,2,4-trimethyl-1,3-pentanediol mono isobutyrate (widely used as plasticizers), propylene glycol and propylene glycol ethers (both used in paints) than to microbial sources ($P \leq 0.003$). Other MVOCs such as terpenes and sesquiterpenes can also be emitted from both anthropogenic or non-microbial sources (fruits, cleaning products, cosmetics, woods, etc.) and from microbes (Dunkel, et al. 2009). The same occurs with methyl-furanes, which apart from being produced by microbes, they are also emitted during the pyrolysis of tobacco components (Schleibinger et al., 2008). There are also several other MVOCs that are related to VOCs liberated from building materials as well (Kuske et al., 2005).

Based on a critical review of the literature over the last decade and existing MVOCs metabolic databases such as the *mVOC* from Lemfack et al. (2014) and the Kyoto Encyclopedia of Genes and Genomes (KEGG) (Kanehisa Laboratories, 2016), a list of MVOCs potentially specific to a microbial species is summarized in Table 6.4. The chemical properties of these MVOCs and the bacterial or fungal producer species are detailed in this table. These MVOCs are less often reported in indoor and outdoor studies but could be an approach for the identification of microbial species by their identification.

6.3   SAMPLING COLLECTION DEVICES

Electronic noses, activated charcoal pads, thermal desorption (TD) tubes, cyclones and glass impingers are the most used techniques for sampling MVOCs (Table 6.5). TD tubes are preferred among the rest of the techniques as they are directly loaded into the GC-MS without needing sample preparation. MVOCs collected by activated charcoal pads, cyclones and glass impingers, for instance, need additional extraction steps. Electronic noses provide a real time analysis of MVOCs, but the sensitivity is not good enough for the detection of these compounds at environmental levels.

There is information available in the literature about which sampling technique should be used to study airborne microorganisms, odors and

TABLE 6.4 Physicochemical Properties of Some Potentially Species-Specific MVOCs

Microbial Species	Specific MVOCs	Molecular formula	Molecular weight (g mol^{-1})	Reference
Aspergillus flavus	cis2-octen-1-ol	$C_8H_{16}O$	128.21	(Morath et al., 2012)
Aspergillus fumigatus	2,4-Pentadione (Acetylacetone)	$C_5H_8O_2$	100.12	(Morath et al., 2012)
	3-Methyl-1,3-pentandione	$CH_3COCH(CH_3)COCH_3$	114.14	
	p-Mentha-6,8-dien-2-ol acetate	$C_{12}H_{18}O_2$	194.27	
Aspergillus versicolor	Trimethylnonanoic acid methylester	$C_{14}H_{28}O_2$	228.37	(Matysik et al., 2009)
	1-(3-Methylphenyl)-ethanone	$C_9H_{10}O$	134.18	
Aspergillus candidus	3-Cyclohepten-1-one isomer	$C_{18}H_{32}O_2$	280.45	(Fischer, et al. 1998)
Emericella nidulans	beta-Fenchyl alcohol	$C_{10}H_{18}O$	154.25	(Fischer, et al. 1998)
	2-Methyl-butanoic acid methyl ester	$C_7H_{14}O_2$	130.18	(Fischer, et al. 1998)
	4,4-Dimethyl-pentenoic acid methyl ester	n/a	n/a	
Penicillium clavigerum	Bicyclooctan-2-one	$C_8H_{12}O$	124.18	(Fischer, et al. 1998)
Penicillium crustosum	2-Ethyl-5-methyl-furan	$C_7H_{10}O$	110.15	(Fischer, et al. 1998)
	4-Ethylbutan-4-olide ((S)-gamma-hexalactone)	$C_6H_{10}O_2$	114.14	(Fischer, et al. 1998)
	Isopropylfuran	$C_7H_{10}O$	110.15	
Penicillium cyclopium	2-Methyl-2-bornene isomer	n/a	n/a	(Fischer, et al. 1998)
	delta-2-Dodecanol	n/a	n/a	
	4-Methyl-2-(3-methyl-2-butenyl)-furan	$C_{10}H_{14}O$	150.22	
Penicillium roqueforti	beta-patchoulene-isomer	$C_{15}H_{24}$	204.35	(Morath et al., 2012)
	beta-elemene-isomer	$C_{15}H_{24}$	204.36	

TABLE 6.4 (Continued)

Microbial Species	Specific MVOCs	Molecular formula	Molecular weight (g mol^{-1})	Reference
	(1,1-dimethylethyl)-2-methylphenol	n/a	n/a	
	Butanoic acid, 2-methyl-2-methylpropyl ester	$C_8H_{16}O_2$	144.21	
	alpha-selinene	$C_{15}H_{24}$	204.35	
	1-methyl-4-(1-methylethyl) benzene (p-Cymene)	$C_{10}H_{14}$	134.22	
	Propanoic acid 2-methyl-2-methylpropyl ester (or Propanoic acid, 2-methyl-3-methylbutyl ester or Isobutyric acid)	$C_8H_{16}O_2$	144.21	
	alpha-chamigrene	$C_{15}H_{24}$	204.35	
Paecilomy cesvariotii	3,5,7-Trimethyl-2E,4E,8E-decatetraene	$C_{13}H_{20}$	176.3	(Pankhurst et al., 2012)
	2-Methyl-2,4-hexadiene	C_7H_{12}	96.17	
	delta-4-Carene	$C_{10}H_{16}$	136.23	
Trichodema pseudokoningii	2-Methyl-pentane	C_6H_{14}	86.18	(Pankhurst et al., 2012)
Muscodor crispans	Hexane, 2,3-dimethyl-	C_8H_{18}	114.23	(Pankhurst et al., 2012)
	Formamide, N-(1-methylpropyl)	$C_7H_{15}NO_2$	145.2	
	Cyclohexane, 1,2-dimethyl-3,5- bis (1-methylethenyl)	$C_{14}H_{24}$	192.34	
Arthrobacter globiformis	2-Phenylethylamine	$C_8H_{11}N$	121.18	(Pankhurst et al., 2012)
Mycobacterium	5-Methylhexan-3-ol	$C_7H_{16}O$	116.2	(Pankhurst et al., 2012)
	7-Methyloctan-3-one	$C_9H_{18}O$	142.24	(Pankhurst et al., 2012)

TABLE 6.4 *(Continued)*

Microbial Species	Specific MVOCs	Molecular formula	Molecular weight (g mol⁻¹)	Reference
	5-Methyl-4-hexen-3-one	$C_7H_{12}O$	112.17	
	Cyanoisoquinoline	$C_{10}H6N_2$	154.17	
Bacillus spp.	(2R,3R)-Butane-2,3-diol	$C_4H_{10}O_2$	90.12	(Pankhurst et al., 2012)
Geobacillus stearothermophillus	Dimethyl ditelluride	C_2H_6Te	157.67	(Pankhurst et al., 2012)
	Methanetellurol	CH_4Te	143.64	
	dimethylselenodisulfide	n/a	173.15	
	dimethyltellurenalsulfide	n/a	189.73	
Paenibacillus polymyxa	2-(2-Methylpropyl)pyrazine	$C_9H_{14}N_2O$	166.22	(Pankhurst et al., 2012)
	2,6-Diisobutylpyrazine	$C_{12}H_{20}N_2$	192.3	
	2-Methyl-5-isobutylpyrazine	$C_9H_{14}N_2$	150.22	
Stapahylococcus aureus	2,3,4,5- tetrahydropyridazine	$C_4H_8N_2$	84.12	(Pankhurst et al., 2012)
	4-methylhexanoic acid	$CH_3CH_2CH(CH_3)$ CH_2CH_2COOH	130.18	
	Butyl butanoate (butyl butyrate)	$C_8H_{16}O_2$	144.21	
Pseudomonas sp.	2,4-Diacetylphloroglucinol	$C_{10}H_{10}O_5$	210.18	(Macnaughton et al., 1997)
Pseudomonas trivialis	Undecadiene	$C_{11}H_{20}$	152.28	(Mette et al., 2015)
	Benzyloxybenzonitrile	$C_{14}H_{11}NO$	209.25	
Escherichia coli	Pentylcyclopropane	C_8H_{16}	112.21	(Mette et al., 2015)
Acinetobacter calcoaceticus	Sulfoacetaldehyde	$C_2H_4O_4S$	124.12	(Mette et al., 2015)

TABLE 6.4 *(Continued)*

Microbial Species	Specific MVOCs	Molecular formula	Molecular weight (g mol⁻¹)	Reference
Klebsiella sp.	Pentylbutanoate (or pentyl butyrate)	$C_9H_{18}O_2$	158.24	(Mette et al., 2015)
Streptomyces citreus	Dihydroagarofuran (sesquiterpenoid)	$C_{15}H_{26}O$	222.37	(Mette et al., 2015)
	Bicyclogermacrene	$C_{15}H_{24}$	204.35	
	betabourbonene	$C_{15}H_{24}$	204.35	
	delta-elemene	$C_{15}H_{24}$	204.36	
Alternaria alternata	6-Methylheptanol	$C_8H_{18}O$	130.23	(Mette et al., 2015)
Rhizopus stolonifer	1-Octene	C_8H_{16}	112.24	(Mette et al., 2015)
	3-Methyl-3-buten-1-ol	$CH_2=C(CH_3)CH_2CH_2OH$	86.13	

Table modified from García-Alcega et al. (2016).

TABLE 6.5 Advantages and Limitations of Different Sampling Techniques

Sampling Techniques	Advantages	Limitations	References
Activated charcoal pads	- Each sample can be analyzed more than once - Good capacities for hydrocarbons, esters, ethers, alcohols, ketones, glycol ethers and halogenated hydrocarbons - Allows long sampling time: MVOCs emissions can be monitored over the time (weeks or days) - Cheap, light and easy to use Are operated without electricity	- Poor recoveries for less volatile and reactive compounds (amines, phenols, aldehydes, and unsaturated hydrocarbons) - Long sampling times (>1h to days or weeks) - Low sensitivity Solvent extraction and heat produce VOC degradation products	(Elke et al., 1999; Hung et al., 2015; Matysik et al., 2009; Özden Üzmez et al., 2015; Park et al., 2002)
Impingers	- High sample volume can be collected - MVOCs are collected in a liquid so the microbes do not get dried - Easily portable	- Possibility of losing sample liquid due to high flow rate - More variability in concentrations between replicates because the sampling time is shorter	(Dybwad et al., 2014; Han and Mainelis, 2008; He and Yao, 2011; Langer et al., 2012; Verreault et al., 2011)
Tenax® Desorption tubes	- Quick as sample preparation is no needed - Good recoveries and precision Easily portable	- Each sample can only be analyzed once - 100 times more sensitive than solvent extraction	(Cauchi et al., 2015; Gao et al., 2002; Müller et al., 2004; Ryan and Beaucham, 2013; Van Huffel et al., 2012)

TABLE 6.5 *(Continued)*

Sampling Techniques	Advantages	Limitations	References
Activated charcoal Desorption tubes	- Collection of very volatile MVOC - Easily portable	- Each sample can only be analyzed once - Poor recoveries for less volatile and reactive compounds (amines, phenols, aldehydes, and unsaturated hydrocarbons) - 100 times more sensitive than solvent extraction	(Matysik et al., 2009; Ström et al., 1994)
Sep-Pak cartridges	- Good for sampling carbonyl compounds	- Not good recoveries for rest of MVOCs	(Korpi et al., 2009)
Electronic nose	- Real time analysis - Portable - User friendly	- Limited detection and identification of MVOCs in the environment - Miss classification and false positive results - Sample pre-concentration and clean up to remove the interferences - Only recognizes compounds present in the library or in the pattern recognition software	(Betancourt et al. 2013; Hurst et al., 2005; Kuske et al., 2005; Morath et al., 2012; Suchorab et al. 2015; Wilson and Baietto, 2011)

Table modified from García-Alcega et al. (2016).

MVOCs at composting facilities and or its surroundings (Albrecht et al., 2008) but there are still lacking guidelines for sampling in other environments like a farm or a park (García-Alcega et al., 2016). Monitoring bioaerosols is difficult because there are several factors that interfere in the sample capture such as the type of activity that is being carried out on the site (turning the compost, incorporation of green waste, no activity at all) as well as the weather conditions (wind speed and wind direction, temperature, relative humidity and atmospheric conditions) or the distance from the emission source. For this reason, a recent guideline from the Environment Agency for sampling bioaerosols from composting sites suggest to collect samples upwind, downwind and at the nearest sensitive receptor in order to be able to compare the different concentrations and also to monitor the weather parameters with a weather station (Environment Agency, 2017). When interpreting the MVOCs data all the weather variables and site activities need to be taken into consideration via multivariate analysis to understand the MVOCs patterns concentrations accordingly.

6.3.1   ELECTRONIC NOSE

Electronic noses function by electronic chemical sensors combined with an information processing unit and pattern recognition software. Volatile organic compounds (VOCs) are then qualitatively recognized from a reference library (Wilson and Baietto, 2009, 2011), which can be constructed for each specific category of microbes (fungi and bacteria) (Wilson et al., 2004). Because advantageously these devices are portable and identify MVOCs in real time, they have often been used to detect fungal contamination in indoor environments as well as in food production processes, agriculture and pathological and clinical diagnoses (Kuske et al., 2005; Wilson, 2013; Wilson and Baietto, 2011). Unfortunately, this technique is not suitable yet for the detection and identification of MVOCs at trace levels because electronic noses are not very sensitive and the sensors cannot discriminate between structurally similar compounds. Moreover, the sensor can easily be activated by random noise and give false identifications or false-positive results. Another inconvenient of this device is the inability of identifying VOCs from complex mixtures where interferences need to be removed by clean-up and pre-concentration steps, which involve a potential loss of sample (Kuske et al., 2005).

6.3.2 ACTIVATED CHARCOAL PADS

Charcoal pads (Figure 6.1) are light diffusive samplers, which are available at low price and are easy to use as they work without electricity (Özden Üzmez et al., 2015). There are not many studies in the literature using charcoal pads for MVOCs analysis. This technique is advantageous for monitoring an area during long times (from hours to weeks) as the device can be left in the field without the need of controlling it (Matysik et al., 2009; Özden Üzmez et al., 2015). Sample preparation is easy as MVOCs are solvent extracted with carbon disulfide (Matysik et al., 2009), but there is a probability of losing sample and, in fact, the sensitivity of charcoal pads is low. Charcoal pads should be stored in the fridge prior analysis and samples should be analyzed within a month (Özden Üzmez et al., 2015).

FIGURE 6.1   Charcoal pad passive sampler.

6.3.3 CYCLONES AND GLASS IMPINGERS

Cyclones and glass impinger sampling devices collect the air sample by suction into a liquid medium avoiding the loss of microbes due to impaction and dryness (Verreault et al., 2011). Another of the advantages of impingers and cyclones is that they can collect a high volume of sample during a relatively short sampling time (10–30 min). However, studies have shown that 30 min sampling time can result in 10 and 15% loss of the collection fluid when using cyclones and glass impingers respectively (Willeke et al., 1998). When considering outdoor sampling, cyclone devices are better suited as they are easier to transport and sterilize than glass impingers. They are also less fragile, and there is no risk of having broken glass. There are a great variety of cyclone samplers used to monitor bioaerosols including the SpinCon air sampler (Specter Industries, Inc.), the BioGuardian air sampler (InnovaTek, Inc.), the BioCapture 650 (MesoSystems Technology, Inc.) and Coriolis®µ (Figure 6.2) (Langer et al., 2012). The advantage of the Coriolis®µ sampler among the other cyclones is its lighter weight (3 kg) and the high sample volume (20 mL). The SpinCon sampler is quite heavy (20 kg), and it operates at a high flow rate (400–450 mL min⁻¹). The BioGuardian air sampler is lighter (7.7 kg), but it works at a low flow rate (90 mL min⁻¹). The Biocapture is also quite light (4–4.5 kg), but the collection volume is small (2–7 mL). One of the most widely used glass impingers is the BioSampler (SKC, Inc.) (Mandal and Brandl, 2011) and there are few studies using the AGI-30 (Ace Glass Inc.). This last one, for instance, is not so efficient collecting samples at shorter time as the BioSampler and also has problems of sample loss at high and low temperatures (Clauß et al., 2011). Liquid samples collected from cyclones and glass impingers should be kept at –20°C until use and analysis should be carried out within a month.

FIGURE 6.2   Coriolis®µ cyclone.

6.3.4   THERMAL DESORPTION TUBES

Thermal desorption (TD) tubes (Figure 6.3) are the most widely used technique for sampling MVOCs in outdoor environments. The advantage of this technique over the others is that it does not require sample preparation, only the addition of an internal standard, often Toluene-d⁸ (Cauchi et al., 2015). Since TD tubes are directly desorbed into the GC-MS, there is no sample loss and allows the detection of outdoor MVOCs at pg level and obtaining good recoveries (Claeson et al., 2002; Müller et al., 2004; Ryan and Beaucham, 2013). TD tubes coated with Tenax® or Tenax®-Carbotrap 50/50 v/v are the most common sorbents (Claeson et al., 2002) although MVOCs are more efficiently collected with a multi-sorbent bed of carbonaceous adsorbents rather than with solely Tenax® (Gallego et al., 2010) (Figure 6.4). Longer sampling times can be applied without break-through of the tubes when using activated charcoal (Anasorb®) coatings, but these are only effective for the collection of very volatile MVOCs, obtaining poor recoveries of aldehydes, unsaturated hydrocarbons, phenols and amines, i.e., the reactive and less volatile MVOCs (Matysik et al., 2009; Ström et al., 1994). Carbonyl compounds are better sampled with TD coated with Sep-Pak® but poor recoveries are obtained for the rest of the MVOCs (Korpi et al., 1998; Korpi et al., 2009). TD tubes should be kept at 4°C before analysis for one month, but the caps of the tubes need to be tightened up after being 1 h in the fridge as they can become loose and samples can be lost.

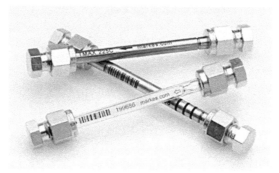

FIGURE 6.3   Thermal desorption tubes.

6.4   ANALYTICAL TECHNIQUES AND CHEMOMETRICS

6.4.1   CHROMATOGRAPHIC TECHNIQUES

The most suitable technique for the analysis of MVOCs is gas chroma-tography coupled to mass spectrometry (GC-MS) due to the volatile nature of these chemicals (Siddiquee et al., 2015). Moreover, trace levels

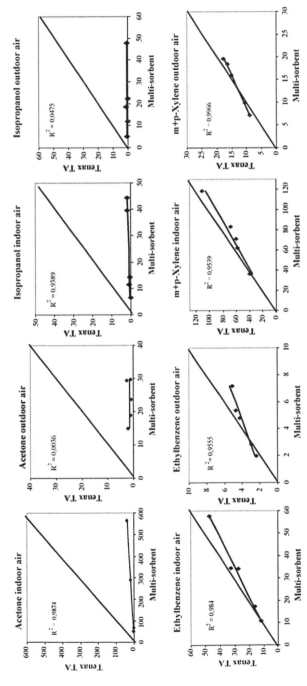

FIGURE 6.4 Comparison of different compound concentrations (µg m⁻³) using multi-sorbent bed tubes (Carbotrap, Carbopack X and Carboxen 569) and Tenax TA tubes. Printed with permission (Gallego et al., 2010).

of MVOCs from complex environmental matrices can be identified and quantified by GC-MS due to its low limit of detection (LOD) (pg m^{-3}) and good chromatographic separation. Different sample injection needs to be performed depending on the sampling accessory that is used (e.g., solvent injection, SPME or TD-GC-MS). Few studies have published MVOCs results using liquid chromatographic techniques (LC-MS), but this is only efficient for carbonyl MVOCs such as aldehydes, ketones, carboxylic acids, carboxylic esters and amines (Korpi et al., 1998) and nonvolatile microbial compounds (Vishwanath et al., 2011).

MVOCs are solvent extracted from charcoal pads, and the extracts are injected into the GC-MS. The main limitation of this technique is that VOCs are degraded during the solvent extraction, and the heat (Matysik et al., 2009) and MVOCs do not have good recoveries due to the sample loss.

TD tubes, for instance, are more appropriate as they are directly loaded into the thermal desorber and connected to the GC-MS without producing any loss of the volatile compounds (Rodríguez-Navas, 2012). When air samples are collected into liquids, headspace solid phase micro-extraction (HS-SPME) is the most common and established MVOC extraction method allowing detection at trace levels (pg L^{-1}) (Wady et al., 2003). The extraction of MVOCs from the liquid sample occurs in a short fused silica fiber without solvents and then they are desorbed at high temperatures and analyzed into the GC-MS. Although this is a robust technique and it offers a high separation capacity, the extraction efficiency is limited to some MVOCs (Morath et al., 2012; Matysik et al., 2009). HS techniques need an SPME fiber in the injector to concentrate the MVOCs even though they are not as sensitive as TD-GC-MS. The advantages and limitations of the different analytical techniques are detailed in Table 6.6.

6.4.2   SPECTROMETRIC TECHNIQUES

There are few studies using FT-IR for analysis of ambient VOCs. This technique measures the frequencies at which the compounds absorb the IR radiation and the intensities of this absorption (Simonescu, 2012). There is an EPA database available with the infrared spectral information of the compounds. The advantages of the FT-IR are its low cost, short analysis time and the possibility of real time analysis. The equipment can be coupled with a thermal desorber so it can analyze TD tubes. The

TABLE 6.6 Advantages and Limitations of Different Analytical Techniques

	Advantages	Limitations	References
GC-MS	- Very good separation - High sensitivity	- Relatively non-portable and slow analysis; - Depending on the source used (Electron ionization or Chemical ionization) some compound can be detected (i.e., 2-methyl-1–butanol is not detectable when EI used; 2-pentanol is not detectable when CI used)	(Došen et al., 2016; Matysik et al., 2009; Morath et al., 2012; Siddiquee et al., 2015; Wady et al., 2003)
GC-HS-SPME-MS	- High sensitivity at trace levels (pg·L⁻¹ to ng·L⁻¹) - Robust and reproducible analytical method - Not sensitive to matrix effects - No solvent is needed - No sample preparation is needed - Powerful separation capacity	- Limited extraction efficiency and some MVOCs might not be detected	(Demeestere et al., 2007; Morath et al., 2012; Sun et al. 2014; Wady et al., 2003; Zhang and Li, 2010; Zhang et al., 2016)
LC-MS	- Good for analysis of carbonyl compounds	- No suitable for volatile compounds	(Cohen et al., 2004; Kildgaard et al., 2014; Korpi et al. 1998)
FT-IR and FT-NIRS	- Low cost - Fast analysis - Environmental friendly	- Low reproducibility and repeatability - Interferences from H_2O, CO_2, and temperature	(Lampert, 2013; Quintelas et al., 2015; Simonescu, 2012)

Table modified from García-Alcega et al. (2016).

main disadvantages are its low reproducibility between samples and the interferences from H_2O and CO_2 in the ambient. There are special traps, but they cannot fully solve this problem. For samples with a large amount of compounds, the data analysis can be complicated as a single compound can have more than one peak (Lampert, 2013).

Fourier Transform Near Infrared Spectroscopy (FT-NIRS) has been used to detect bacterial contamination in water based pharmaceutical preparations (Quintelas et al., 2015). To the author's knowledge this technique has not been used to study MVOCs yet and could be explored as it provides a very fast analysis, is environmentally friendly and low cost. Disadvantageously, it has the same limitations as FT-IR, the low reproducibility, and repeatability due to the temperature, water and CO_2 interferences (Quintelas et al., 2015).

Since different compounds require the use of different ionization sources, for example, the identification of 2-pentanol can only be achieved by electron impact ionization (EI) and 2-methyl-1-butanol by chemical impact (CI) (Matysik et al., 2009; Wady et al., 2003), it would be recommendable to use a combination of both EI and CI to be able to scan a wider range of MVOCs in the environment.

6.4.3   USING CHEMOMETRICS FOR ANALYZING MVOCS PROFILES

Chemometrics is the mathematical and statistical analysis of the chemical data obtained from the chromatogram of a sample. By chemometrics, the maximum information about the compounds of study is extracted by optimizing signal and data analysis processes and performing multivariate analysis in order to study chemical trends (Vivó-Truyols et al., 2005). The use of this approach to identify microorganisms that are in the air is receiving increased attention as chemometrics is a cost effective and fast analysis in comparison with the more traditional molecular or cell culturing techniques (Lemfack et al., 2014).

The schematic procedure for the analysis of the samples is represented in Figure 6.5. The procedure of the analysis of MVOCs from environmental air starts with the noise removal of the chromatograms (using SNIP baseline detector for example) followed by peak deconvolution, which allows accurate mass spectra identification from complex chromatograms

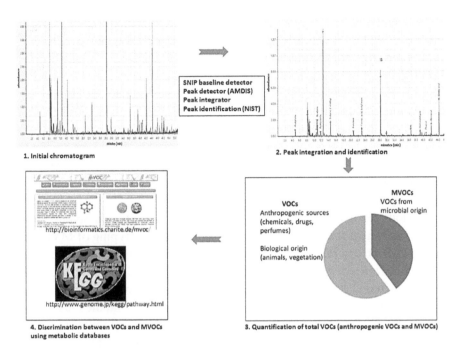

1. Initial chromatogram

2. Peak integration and identification

SNIP baseline detector
Peak detector (AMDIS)
Peak integrator
Peak identification (NIST)

http://bioinformatics.charite.de/mvoc

http://www.genome.jp/kegg/pathway.html

4. Discrimination between VOCs and MVOCs
using metabolic databases

VOCs
Anthropogenic sources
(chemicals, drugs,
perfumes)

Biological origin
(animals, vegetation)

MVOCs
VOCs from
microbial origin

3. Quantification of total VOCs (anthropogenic VOCs and MVOCs)

FIGURE 6.5 **(See color insert.)** Treatment of a chromatogram from a WWTP air sample for peak identification and separation of MVOCs from VOCs with online databases.

(e.g., AMDIS). Then, it proceeds with the identification of all the peaks within the chromatograms of the samples, and this can be done with NIST or free databases such as mzCloud or METLIN. The mzCloud uses a new third generation spectra correlation algorithm to search and identify the compounds. METLIN is a metabolomics database useful for the identification of metabolites, which are linked to the KEGG database (Kanehisa Laboratories, 2016) to see the metabolic pathways and the microorganisms producers. The MVOCs analysis for ambient air is complex because not all MVOCs have solely microbial origin; these also can be anthropogenic or produced by fruits or vegetables. To discriminate MVOCs from VOCs there is an approach available looking at metabolical databases such as KEGG database (Kanehisa Laboratories, 2016) or *mVOC* database (Lemfack et al., 2014). After identifying the chromatogram peaks, the *m/z* spectra of the MVOCs are analyzed statistically by multivariate analysis (Schenkel et al., 2015), hierarchical Cluster Analysis (HCA) or multidimensional scaling analysis (MDS) and/or principal component analysis (PCA) to study the

MVOCs patterns and trends across the samples (Schenkel et al., 2015; Sun et al., 2014). These analysis can be done using chemometric software such as among others SpectConnect, ACD/MS Manager, OpenChrom, Mass Profiler Professional, or commercial statistical software (MATLAB, ADAPT, etc.) (Murphy et al., 2012). These statistical analyses help us to identify the key and more representative compounds per site, giving us a preliminary idea of the potential markers for each site. Then, the microbial identity of these potentially species-specific markers should be verified by correlation with DNA sequencing analysis. Specific microbial markers could be used in the future for the identification of microbes in air ideally with a sampling device that gives real time data.

Previous researchers have demonstrated the feasibility and potential of identifying fungal species by chemotaxonomy or chemotyping of MVOCs emissions from *in vitro* cultures (Hung and Bennett, 2015; Hussain et al., 2010; Polizzi et al., 2009; Wihlborg, et al., 2008). Wihlborg et al. (2008) cultured different fungi species and detected 118 different MVOCs. The PCAs obtained from the MVOCs profiles demonstrated that the clustering of the various fungi types depends more on the taxonomy rather than on the culturing medium although this also affected the MVOCs production. Following this work, more emphasis needs to be done towards the development of a species-specific data set containing MVOC and their concentrations from outdoor environments such as industrial (composting facilities, WWTPs, waste facilities), urban (parks, neighborhoods) and rural (farms, countryside) areas.

6.5   CONCLUSIONS AND FUTURE PERSPECTIVES

Although bioaerosols have been studied over the last 30 years, the understanding of their composition is still in its infancy, and further inputs are needed to improve the identification, characterization, and quantification of bioaerosols emitted from urban, rural and industrial environments. Ambient bioaerosol characterization looking at MVOCs biomarkers seems to be a fast and a reproducible approach as current analytical techniques are more sensitive and allow the detection of chemicals at environmental concentrations. There are numerous variables interfering in the capture of bioaerosols such as weather conditions (wind speed and wind direction, temperature, relative humidity, and atmospheric conditions) and activity that is being carried out on the site. When interpreting the MVOCs data these variables need to be taken into consideration via multivariate analysis in order to understand the MVOCs patterns concentrations.

Currently, there is a lack of standardizing methodology for collecting and analyzing MVOCs in a rapid, reliable and reproducible way. Analytical techniques offer a faster and more economical analysis compared to the traditional molecular and microbial ways but the techniques available at the moment are laboratory based, and the data acquisition is not in real time. TD sampling tubes coupled with GC-MS analysis appears to be the most widely used approach as well as it is also a very sensitive and robust technique but the analysis cannot be offered in real time. To obtain accurate real time outdoor MVOCs measurements, an upgraded electronic nose with smaller size, improved and more specific sensors and algorithms for detection and identification of these chemicals at trace levels should be developed.

The next steps for the identification of species specific MVOCs involve both, the chemical characterization of bioaerosols and the speciation analysis simultaneously. Forthcoming directions should point towards the progression of the development of a rapid analysis and assessment of different outdoor air from rural, industrial and urban environments with a combination of a powerful electronic nose and a standardized database with a list of species-specific MVOCs from contrasting outdoor environments. The aim of this database would be to accelerate the process of identifying pathogen microorganisms in the air without molecular or cell culture techniques.

KEYWORDS

- **analytical techniques**
- **bioaerosols**
- **microbial**
- **MVOCs**
- **samplig collection**
- **VOCs**

REFERENCES

Albrecht, A., Fischer, G., Brunnemann-Stubbe, G., Jäckel, U., & Kämpfer, P., (2008). Recommendations for study design and sampling strategies for airborne microorganisms, MVOC and odours in the surrounding of composting facilities. *Int. J. Hyg. Environ. Health.*, *211*, 121–131.

Araki, A., Kanazawa, A., Kawai, T., Eitaki, Y., Morimoto, K., Nakayama, K., Shibata, E., Tanaka, M., Takigawa, T., Yoshimura, T., Chikara, H., Saijo, Y., & Kishi, R., (2012). The relationship between exposure to microbial volatile organic compound and allergy prevalence in single-family homes. *Sci. Total Environ.*, *423*, 18–26.

Betancourt, D. A., Krebs, K., Moore, S. A., & Martin, S. M., (2013). Microbial volatile organic compound emissions from *stachybotrys chartarum* growing on gypsum wallboard and ceiling tile. *BMC Microbiol.*, *13*, 1–10.

Bos, L. D. J., Sterk, P. J., & Schultz, M. J., (2013). Volatile metabolites of pathogens: A systematic review. *PLoS Pathog.*, *9*, 1–8.

Cauchi, M., Fowler, D. P., Walton, C., Turner, C., Waring, R. H., Ramsden, D. B., Hunter, J. O., Teale, P., Cole, J. A., & Bessant, C., (2015). Comparison of GC-MS, HPLC-MS and SIFT-MS in conjunction with multivariate classification for the diagnosis of Crohn's disease in urine. *Anal. Methods.*, *7*, 8379–8385.

Chambers, S. T., Bhandari, S., Scott-Thomas, A., & Syhre, M., (2011). Novel diagnostics: progress toward a breath test for invasive *Aspergillus fumigatus*. *Med. Mycol.*, *49*, 54–61.

Choi, H., Schmidbauer, N., & Bornehag, C. G., (2016). Non-microbial sources of microbial volatile organic compounds. *Environ. Res.*, *148*, 127–136.

Claeson, A., Levin, J., Blomquist, G., & Sunesson, A., (2002). Volatile metabolites from microorganisms grown on humid building materials and synthetic media. *J. Env. Monit.*, *4*, 667–672.

Clauss, A., Clauss, M., & Hartung, J., (2011). A temperature-controlled AGI-30 impinger for sampling of *bioaerosols*. *Aerosol Sci. Technol.*, *45*, 1231–1239.

Cohen, D. D., Garton, D., Stelcer, E., Hawas, O., Wang, T., Poon, S., Kim, J., Choi, B. C., Oh, S. N., & Shin, H. J., (2004). Multielemental analysis and characterization of fine aerosols at several key ACE-Asia sites. *J. Geophys. Res. D-Atmospheres.*, *109*, D19S12.

Demeestere, K., Dewulf, J., De Witte, B., & Van Langenhove, H., (2007). Sample preparation for the analysis of volatile organic compounds in air and water matrices. *J. Chromatogr. A.*, *1153*, 130–144.

Došen, I., Nielsen, K. F., Clausen, G., & Andersen, B., (2016). Potentially harmful secondary metabolites produced by indoor *chaetomium* species on artificially and naturally contaminated building materials. *Indoor Air.*, *27*, 1–13.

Dunkel, M., Schmidt, U., Struck, S., Berger, L., Gruening, B., Hossbach, J., Jaeger, I. S., Effmert, U., Piechulla, B., & Eriksson, R., (2009). SuperScent–a database of flavors and scents. *Nucleic Acids Res.*, *37*, 291–294.

Dybwad, M., Skogan, G., & Blatny, J. M., (2014). Comparative testing and evaluation of nine different air samplers: end-to-end sampling efficiencies as specific performance measurements for bioaerosol applications. *Aerosol. Sci. Technol.*, *48*, 282–295.

Eiceman, G. A., & Stone, J. A., (2004). Ion mobility spectrometry in homeland security. *Anal. Chem.*, *76*, 390–397.

Elke, K., Begerow, J., Oppermann, H., Krämer, U., Jermann, E., & Dunemann, L., (1999). Determination of selected microbial volatile organic compounds by diffusive sampling and dual-column capillary GC-FID - A new feasible approach for the detection of an exposure to indoor mold fungi? *J. Environ. Monit.*, *1*, 445–452.

Environment Agency, UK Government. M9: Environmental monitoring of bioaerosols at regulated facilities. LIT 10645, 2017.

Fischer, G., Albrecht, A., Jäckel, U., & Kämpfer, P., (2008). Analysis of airborne microorganisms, MVOC and odor in the surrounding of composting facilities and implications for future investigations. *Int. J. Hyg. Environ. Health.*, *211*, 132–142.

Fischer, G., Muller, T., Thissen, R., Braun, S., & Dott, W., (2004). Process-dependent emission of airborne fungi and MVOC from composting facilities. *Gefahrst. Reinhalt. Luft.*, *64*, 160–167.

Fischer, G., Schwalbe, R., Ostrowski, R., & Dott, W., (1998). Airborne fungi and their secondary metabolites in working places in a compost facility. *Mycoses.*, *41*, 383–388.

Gallego, E., Roca, F. J., Perales, J. F., & Guardino, X., (2010). Comparative study of the adsorption performance of a multi-sorbent bed (Carbotrap, Carbopack X, Carboxen 569) and a Tenax TA adsorbent tube for the analysis of volatile organic compounds (VOCs). *Talanta.*, *81*, 916–924.

Gao, P., Korley, F., Martin, J., & Chen, B. T., (2002). Determination of unique microbial volatile organic compounds produced by five Aspergillus species commonly found in problem buildings. *Am. Ind. Hyg. Assoc. J.*, *63*, 135–140.

Garcia-Alcega, S., Nasir, Z. A., Ferguson, R., Withby, C., Dumbrell, A. J., Colbeck, I., Gomes, D., Tyrrel, S., & Coulon, F., (2016). Fingerprinting outdoor air environment using microbial volatile organic compounds (MVOCs) - A review. *Trends Anal. Chem.*, *86*, 75–83.

Ghosh, B., Lal, H., & Srivastava, A., (2015). Review of bioaerosols in indoor environment with special reference to sampling, analysis and control mechanisms. *Environ. Int.*, *85*, 254–272.

Gómez-Domenech, M., García-Mozo, H., Alcázar, P., Brandao, R., Caeiro, E., Munhoz, V., & Galán, C., (2010). Evaluation of efficiency of the Coriolis air sampler for pollen detection in south Europe. *Aerobiologia (Bologna)*, *26*, 149–155.

Han, T., & Mainelis, G., (2008). Design and development of an electrostatic sampler for bioaerosols with high concentration rate. *J. Aerosol. Sci.*, *39*, 1066–1078.

He, Q., & Yao, M., (2011). Integration of high volume portable aerosol-to-hydrosol sampling and qPCR in monitoring bioaerosols. *J. Environ. Monit.*, *13*, 706–712.

Hung, R., Lee, S., & Bennett, J. W., (2015). Fungal volatile organic compounds and their role in ecosystems. *Appl. Microbiol. Biotechnol.*, *99*, 3395–3405.

Hurst, C., Longhurst, P., Pollard, S., Smith, R., Jefferson, B., & Gronow, J., (2005). Assessment of municipal waste compost as a daily cover material for odor control at landfill sites. *Environ. Pollut.*, *135*, 171–177.

Hussain, A., Tian, M. Y., He, Y. R., & Lei, Y. Y., (2010). Differential fluctuation in virulence and VOC profiles among different cultures of entomopathogenic fungi. *J. Invertebr. Pathol.*, *104*, 166–171.

Kanehisa Laboratories, (2016). KEGG Pathway Database. http://www. kegg. jp/kegg/ pathway. html (accessed Apr. 20, 2016).

Kildgaard, S., Mansson, M., Dosen, I., Klitgaard, A., Frisvad, J. C., Larsen, T. O., & Nielsen, K. F., (2014). Accurate dereplication of bioactive secondary metabolites from marine-derived fungi by UHPLC-DAD-QTOFMS and a MS/HRMS library. *Mar. Drugs., 12,* 3681–3705.

Kim, J., Jin, J. H., Kim, H. S., Song, W., Shin, S. K., Yi, H., Jang, D. H., Shin, S., & Lee, B. Y., (2016). Fully automated field-deployable bioaerosol monitoring system using carbon nanotube-based biosensors. *Environ. Sci. Technol., 50,* 5163–5171.

Kim, Y., Platt, U., Gu, M. B., & Iwahashi, H., (2009). *Atmospheric and Biological Environmental Monitoring,* Springer Science+Business Media B.V., eBook ISBN 978-1-4020-9674-7, pp. 311.

Konuma, R., Umezawa, K., Mizukoshi, A., Kawarada, K., & Yoshida, M., (2015). Analysis of microbial volatile organic compounds produced by wood-decay fungi. *Biotechnol. Lett., 37,* 1845–1852.

Korpi, A., Jarnberg, J., & Pasanen, A. L., (2009). Microbial volatile organic compounds. *Crit. Rev. Toxicol., 39,* 139–193.

Korpi, A., Pasanen, A. L., & Pasanen, P., (1998). Volatile compounds originating from mixed microbial cultures on building materials under various humidity conditions. *Appl. Environ. Microbiol., 64,* 2914–2919.

Kuske, M., Romain, A. C., & Nicolas, J., (2005). Microbial volatile organic compounds as indicators of fungi. Can an electronic nose detect fungi in indoor environments? *Build. Environ., 40,* 824–831.

Lampert, C., (2013). Analysis of volatile organic compounds in air samples by infrared spectroscopic analysis of sorbent tube samples. Middle Tennessee State University, Murfreesboro (TN), MS Dissertation, 56 pages.

Langer, V., Hartmann, G., Niessner, R., & Seidel, M., (2012). Rapid quantification of bioaerosols containing *L. pneumophila* by Coriolis® μ air sampler and chemiluminescence antibody microarrays. *J. Aerosol. Sci., 48,* 46–55.

Lavine, B. K., Mirjankar, N., LeBouf, R., & Rossner, A., (2012). Prediction of mold contamination from microbial volatile organic compound profiles using solid phase microextraction and gas chromatography/mass spectrometry. *Microchem. J., 103,* 37–41.

Lehtinen, J., & Veijanen, A., (2011). Odor monitoring by combined TD–GC–MS–Sniff technique and dynamic olfactometry at the wastewater treatment plant of low H_2S concentration. *Water Air Soil Pollut., 218,* 185–196.

Lehtinen, J., Tolvanen, O., Nivukoski, U., Veijanen, A., & Hänninen, K., (2013). Occupational hygiene in terms of volatile organic compounds (VOCs) and bioaerosols at two solid waste management plants in Finland. *Waste Manag., 33,* 964–973.

Lemfack, M. C., Nickel, J., Dunkel, M., Preissner, R., & Piechulla, B., (2014). mVOC: a database of microbial volatiles. *Nucleic Acids Res., 42*(Database issue), 744–748.

Lorenz, W., Diederich, T., & Conrad, M., (2002). Practical experiences with MVOC as an indicator for microbial growth. *Proceedings of Indoor Air,* 341–346.

Macklin, Y., Kibble, A., & Pollitt, F., (2011). *Impact on Health of Emissions of Landfill Sites, Advice of Health Protection Agency, Radiation.* London: Chilton, Didcot, Oxfordshire: Health protection agency, Centre for radiation, Chemical and Environmental Hazards, ISBN 978-0-85951-704-1.

Mandal, J., & Brandl, H., (2011). Bioaerosols in indoor environment: A review with special reference to residential and occupational locations. *Open. Environ. Biol. Monit. J.*, *41*, 83–96.

Mason, S., Cortes, D., & Horner, W. E., (2010). Detection of gaseous effluents and by-products of fungal growth that affect environments (RP-1243). *HVAC&R Res.*, *16*, 109–121.

Matysik, S., Herbarth, O., & Mueller, A., (2009). Determination of microbial volatile organic compounds (MVOCs) by passive sampling onto charcoal sorbents. *Chemosphere.*, *76*, 114–119.

Mcnaughton, S. J., Jenkins, T. L., Wimpee, M. H., Cormiér, M. R., & White, D. C., (1997). Rapid extraction of lipid biomarkers from pure culture and environmental samples using pressurized accelerated hot solvent extraction. *J. Microbiol. Methods.*, *31*, 19–27.

Mette, A., Zervas, A., Tendal, K., & Lund, J., (2015). Microbial diversity in bioaerosol samples causing ODTS compared to reference bioaerosol samples as measured using Illumina sequencing and MALDI-TOF. *Environ. Res.*, *140*, 255–267.

Morath, S. U., Hung, R., & Bennett, J. W., (2012). Fungal volatile organic compounds: A review with emphasis on their biotechnological potential. *Fungal. Biol. Rev.*, *26*, 73–83.

Müller, T., Thissen, R., Braun, S., Dott, W., Fischer, G., Muller, T., Thissen, R., Braun, S., Dott, W., & Fisher, G., (2004). (M)VoC and composting facilities Part 1:(M) VOC emissions from municipal biowaste and plant refuse. *Environ. Sci. Pollut. Res.*, *11*, 91–97.

Murphy, K. R., Parcsi, G., & Stuetz, R. M., (2014). Non-methane volatile organic compounds predict odor emitted from five tunnel ventilated broiler sheds. *Chemosphere.*, *95*, 423–432.

Murphy, K. R., Wenig, P., Parcsi, G., Skov, T., & Stuetz, R. M., (2012). Characterizing odorous emissions using new software for identifying peaks in chemometric models of gas chromatography-mass spectrometry datasets. *Chemom. Intell. Lab. Syst.*, *118*, 41–50.

O'Connor, D. J., Daly, S. M., & Sodeau, J. R., (2015). Online monitoring of airborne bioaerosols released from a composting/green waste site. *Waste. Manag.*, *42*, 23–30.

Oliver, J. D., (2005). The viable but nonculturable state in bacteria. *J. Microbiol.*, *43*, 93–100.

Özden Üzmez, Ö., Gaga, E. O., & Döğeroğlu, T., (2015). Development and field validation of a new diffusive sampler for determination of atmospheric volatile organic compounds. *Atmos. Environ.*, *107*, 174–186.

Pankhurst, L. J., Whitby, C., Pawlett, M., Larcombe, L. D., Mckew, B., Deacon, L. J., Morgan, S. L., Villa, R., Drew, G. H., Tyrrel, S., Pollard, S. J. T., & Coulon, F., (2012). Temporal and spatial changes in the microbial bioaerosol communities in green-waste composting. *FEMS Microbiol. Ecol.*, *79*, 229–239.

Pardon, G., Ladhani, L., Sandström, N., Ettori, M., Lobov, G., & Van der Wijngaart, W., (2015). Aerosol sampling using an electrostatic precipitator integrated with a microfluidic interface. *Sens. Actuator. B-Chem.*, *212*, 344–352.

Park, J. S., & Ikeda, K., (2002). MVOC emissions from fungi in HVAC system. *Indoor Air.*, 335–340.

Polizzi, V., Delmulle, B., Adams, A., Moretti, A., Susca, A., Picco, A. M., Rosseel, Y., Kindt, R., Van Bocxlaer, J., De Kimpe, N., Van Peteghem, C, & De Saeger, S., (2009).

JEM spotlight: Fungi, mycotoxins and microbial volatile organic compounds in moldy interiors from water-damaged buildings. *J. Environ. Monit.*, *11*, 1849–1858.

Prospero, J. M., Blades, E., Mathison, G., & Naidu, R., (2005). Interhemispheric transport of viable fungi and bacteria from Africa to the Caribbean with soil dust. *Aerobiologia (Bologna)*, *21*, 1–19.

Quintelas, C., Mesquita, D. P., Lopes, J. A., Ferreira, E. C., & Sousa, C., (2015). Near-infrared spectroscopy for the detection and quantification of bacterial contaminations in pharmaceutical products. *Int. J. Pharm.*, *492*, 199–206.

Rodríguez-Navas, C., Forteza, R., & Cerdà, V., (2012). Use of thermal desorption-gas chromatography-mass spectrometry (TD-GC-MS) on identification of odorant emission focus by volatile organic compounds characterisation. *Chemosphere.*, *89*, 1426–1436.

Ryan, T. J., & Beaucham, C., (2013). Dominant microbial volatile organic compounds in 23 US homes. *Chemosphere.*, *90*, 977–985.

Sandström, N., Shafagh, R. Z., Vastesson, A., Carlborg, C. F., Van der Wijngaart, W., & Haraldsson, T., (2015). Reaction injection molding and direct covalent bonding of OSTE+ polymer microfluidic devices. *J. Micromech. Microeng.*, *25*, 75002–75014.

Schenkel, D., Lemfack, M. C., Piechulla, B., & Splivallo, R., (2015). A meta-analysis approach for assessing the diversity and specificity of belowground root and microbial volatiles. *Front. Plant. Sci.*, *6*, 707–718.

Schleibinger, H., Laussmann, D., Bornehag, C. G., Eis, D., & Rueden, H., (2008). Microbial volatile organic compounds in the air of moldy and mold-free indoor environments. *Indoor Air.*, *18*, 113–124.

Schmidt, M. S., & Bauer, A. J. R., (2010). Preliminary correlations of feature strength in spark-induced breakdown spectroscopy of bioaerosols with concentrations measured in laboratory analyses. *Appl. Opt.*, *49*, 101–110.

Sharma, A., Clark, E., McGlothlin, J. D., & Mittal, S. K., (2015). Efficiency of airborne sample analysis platform (ASAP) bioaerosol sampler for pathogen detection. *Front. Microbiol.*, *6*, 512.

Siddiquee, S., Al Azad, S., Bakar, F. A., Naher, L., & Kumar, S. V., (2015). Separation and identification of hydrocarbons and other volatile compounds from cultures of Aspergillus niger by GC–MS using two different capillary columns and solvents. *J. Saudi Chem. Soc.*, *19*, 243–256.

Simonescu, C. C. M., (2012). Application of FT-IR spectroscopy in environmental studies. In: *Advanced Aspects of Spectroscopy*, pp. 49.

Srivastava, A., Pitesky, M. E., Steele, P. T., Tobias, H. J., Fergenson, D. P., Horn, J. M., Russell, S. C., Czerwieniec, G. A., Lebrilla, C. B., & Gard, E. E., (2005). Comprehensive assignment of mass spectral signatures from individual *bacillus tropheus* spores in matrix-free laser desorption/ionization bioaerosol mass spectrometry. *Anal. Chem.*, *77*, 3315–3323.

Stotzky, G., & Schenck, S., (1976). Volatile organic compounds and microorganisms. *CRC Crit. Rev. Microbiol.*, *4*, 333–382.

Ström, G., West, J., Wessén, B., & Palmgren, U., (1994). Quantitative analysis of microbial volatiles in damp Swedish houses. In: *Health Implications of Fungi in Indoor Environments*, Samson, R. A., Flannigan, B., Flannigan, M. E., Verhoeff, A. P., Adan, O. C. G., Hoekstra, E. S., ed., Elsevier: North-Holland Biomedical Press, Amsterdam, *1*, pp. 291–305.

Suchorab, Z., Guz, Ł., Łagód, G., & Sobczuk, H., (2015). The possibility of building classification for mold threat using gas sensors array. *Adv. Mater. Res., 1126,* 161–168.

Sun, D., Wood-Jones, A., Wang, W., Vanlangenberg, C., Jones, D., Gower, J., Simmons, P., Baird, R. E., & mLsna, T. E., (2014). Monitoring MVOC profiles over time from isolates of *aspergillus flavus* using SPME GC-MS. *J. Agric. Chem. Environ., 3,* 48–63.

Syhre, M., Scotter, J. M., & Chambers, S. T., (2008). Investigation into the production of 2-pentylfuran by *aspergillus fumigatus* and other respiratory pathogens in vitro and human breath samples. *Med. Mycol., 46,* 209–215.

Toprak, E., & Schnaiter, M., (2013). Fluorescent biological aerosol particles measured with the waveband integrated bioaerosol sensor WIBS-4: laboratory tests combined with a one year field study. *Atmos. Chem. Phys., 13,* 225–243.

Usachev, E. V., Pankova, A. V., Rafailova, E. A., Pyankov, O. V., & Agranovski, I. E., (2012). Portable automatic bioaerosol sampling system for rapid on-site detection of targeted airborne microorganisms. *J. Environ. Monit., 14,* 2739–2745.

Van Huffel, K., Heynderickx, P. M., Dewulf, J., & Van Langenhove, H., (2012). Measurement of odorants in livestock buildings: SIFT-MS and TD-GC-MS. *Chem. Eng. Trans., 30,* 67–72.

Verreault, D., Gendron, L., Rousseau, G. M., Veillette, M., Massé, D., Lindsley, W. G., Moineau, S., & Duchaine, C., (2011). Detection of airborne lactococcal bacteriophages in cheese manufacturing plants. *Appl. Environ. Microbiol., 77,* 491–497.

Vishwanath, V., Sulyok, M., Weingart, G., Kluger, B., Täubel, M., Mayer, S., Schuhmacher, R., Krska, R., Täubel, M., Mayer, S., Schuhmacher, R., & Krska, R., (2011). Evaluation of settled floor dust for the presence of microbial metabolites and volatile anthropogenic chemicals in indoor environments by LC-MS/MS and GC-MS methods. *Talanta, 85,* 2027–2038.

Vivó-Truyols, G., Torres-Lapasió, J. R., Van Nederkassel, A. M., Vander Heyden, Y., & Massart, D. L., (2005). Automatic program for peak detection and deconvolution of multi-overlapped chromatographic signals part I: peak detection. *J. Chromatogr. A., 1096,* 133–145.

Wady, L., Bunte, A., Pehrson, C., & Larsson, L., (2003). Use of gas chromatography-mass spectrometry/solid phase microextraction for the identification of MVOCs from moldy building materials. *J. Microbiol. Methods, 52,* 325–332.

Wihlborg, R., Pippitt, D., & Marsili, R., (2008). Headspace sorptive extraction and GC-TOFMS for the identification of volatile fungal metabolites. *J. Microbiol. Methods., 75,* 244–250.

Willeke, K., Lin, X., & Grinshpun, S. A., (1998). Improved aerosol collection by combined impaction and centrifugal motion. *Aerosol. Sci. Technol., 28,* 439–456.

Willers, C., Jansen van Rensburg, P. J., & Claassens, S., (2015). Phospholipid fatty acid profiling of microbial communities–a review of interpretations and recent applications. *J. Appl. Microbiol., 119,* 1207–1218.

Wilson, A. D., & Baietto, M., (2009). Applications and advances in electronic-nose technologies. *Sensors, 9,* 5099–5148.

Wilson, A. D., & Baietto, M., (2011). Advances in electronic-nose technologies developed for biomedical applications. *Sensors., 11,* 1105–1176.

Wilson, A. D., (2013). Diverse applications of electronic-nose technologies in agriculture and forestry. *Sensors, 13,* 2295–2348.

Wilson, A. D., Lester, D. G., & Oberle, C. S., (2004). Development of conductive polymer analysis for the rapid detection and identification of phytopathogenic Microbes. *Phytopathology, 94,* 419–431.

Yoo, K., Lee, T. K., Choi, E. J., Yang, J., Shukla, S. K., Hwang, S., & Park, J., (2016). Molecular approaches for the detection and monitoring of microbial communities in bioaerosols: A review. *J. Environ. Sci., 51,* 1–15.

Zhang, Q., Zhou, L., Chen, H., Wang, C., & Xia, Z., (2016). Solid-Phase microextraction technology for in vitro and in vivo metabolite analysis. *Trends Anal. Chem., 80,* 57–65.

Zhang, Z., & Li, G., (2010). A review of advances and new developments in the analysis of biological volatile organic compounds. *Microchem. J., 95,* 127–139.

CHAPTER 7

BREATHOMICS AND ITS APPLICATION FOR DISEASE DIAGNOSIS: A REVIEW OF ANALYTICAL TECHNIQUES AND APPROACHES

DAVID J. BEALE,[1] OLIVER A. H. JONES,[2]
AVINASH V. KARPE,[3] DING Y. OH,[4] IAIN R. WHITE,[5]
KONSTANTINOS A. KOUREMENOS,[6] and ENZO A. PALOMBO[3]

[1]Commonwealth Scientific & Industrial Research Organization (CSIRO), Land & Water, P.O. Box 2583, Brisbane, Queensland 4001, Australia, Tel.: +61 7 3833 5774, E-mail: david.beale@csiro.au

[2]Australian Centre for Research on Separation Science, School of Science, RMIT University, PO Box 2547, Melbourne, Victoria 3000, Australia, E-mail: oliver.jones@rmit.edu.au

[3]Department of Chemistry and Biotechnology, Swinburne University of Technology. PO Box 218, Hawthorn, Victoria 3122, Australia, E-mail: akarpe@swin.edu.au, epalombo@swin.edu.au

[4]WHO Collaborating Centre for Reference and Research on Influenza (VIDRL), Peter Doherty Institute for Infection and Immunity, 792 Elizabeth Street, Melbourne, VIC, Australia, and School of Applied and Biomedical Sciences, Federation University, Churchill, Victoria, Australia, E-mail: dingthomas.oh@influenzacentre.org

[5]Manchester Institute of Biotechnology, University of Manchester, Princess St, Manchester M1 7DN United Kingdom, E-mail: iain.white-2@manchester.ac.uk

[6]Metabolomics Australia, Bio21 Molecular Science and Biotechnology Institute, University of Melbourne, 30 Flemington Road, Parkville, VIC, 3010, Australia, E-mail: konstantinos. kouremenos@unimelb.edu.au

ABSTRACT

The application of metabolomics to an ever-greater variety of sample types is a key focus of systems biology research. Recently, there has been a strong focus on applying these approaches toward the rapid analysis of metabolites found in non-invasively acquired samples, such as exhaled breath (also known as 'breathomics'). The sampling process involved in collecting exhaled breath is nonintrusive and comparatively low-cost. It uses a series of globally approved methods and provides researchers with easy access to the metabolites secreted by the human body. Owing to its accuracy and rapid nature, metabolomic analysis of breath is a rapidly growing field that has proven effective in detecting and diagnosing the early stages of numerous diseases and infections. This review discusses the various collection and analysis methods currently applied in breathomics research. Some of the salient research completed in this field to date is also assessed and discussed in order to provide a basis for possible future scientific directions.

7.1   INTRODUCTION

Recent advances in analytical technologies and high-throughput data analysis techniques have established metabolomics as one of the emerging '*omics*' platforms within systems biology research (Beale et al., 2016; Bujak et al., 2015). The latter comprises (meta) genomics, (meta) transcriptomics, (meta)proteomics, and (community) metabolomics (Beale et al., 2017b; Jones et al., 2014). For the purpose of this review, metabolomics is defined as the study of all the small (<1,000 amu) chemical compounds that are either produced or consumed by a biological system. If done correctly metabolomics can provide an unbiased analysis of all such small-molecule metabolites (collectively known as the metabolome) within a given biological system, under a given set of conditions, through the analysis of a variety of different, complementary samples (Snowden et al., 2012).

A recent review by Bujak et al. (2015) about metabolomics-based diagnostics highlighted that breath analysis currently only constitutes a small fraction of the total number of sample matrices investigated in metabolomics. The authors reported that between 2010 and 2015 breath samples were only ~1.4% of metabolomics research, whereas blood (plasma/serum) (68.5%), and urine (21.2%) were the preferred sample matrices, followed by saliva (4.6%), and tissue homogenates (4.3%). Furthermore, collection methods for such samples are generally either considered invasive (such as blood and tissues) or non-invasive (such as saliva, breath, urine, and feces). The cost of sampling and analyzing invasive samples can be relatively high and the collection of some sample types, such as feces and urine, can be socially awkward or embarrassing for the patient. To eliminate such issues, the collection of non-invasive samples such as exhaled breath is increasingly being considered. Furthermore, predefined sample collection protocols and tools that facilitate patient self-sampling are available. As such, this review will focus on the use of non-invasive breath related metabolomics applied to medicine, concentrating on sample collection and analysis approaches. Case study applications of their use in disease diagnosis, a discussion on the future direction of breathomics as a diagnostic medium, and use in personalized medicine are also presented.

7.2   BREATH SAMPLING CONSIDERATIONS

Diagnostic breath testing is familiar to most through roadside testing for alcohol (via fuel cell based sensors) and perhaps to some via the ^{14}C-urea breath test for assessment of gastric *Helicobacter pylori* infection (Goddard and Logan, 1997; Tewari et al., 2001). Breath condensate samples ready for testing with the latter test are shown in Figure 7.1.

With the advancement of metabolomics technology, however, exhaled breath can be leveraged as a far more advanced clinical and diagnostic tool due to its potential to mirror respiratory health, as well as other systemic health conditions within the human body (Cao and Duran, 2006). Breath offers an attractive sample alternative to blood as the collection is relatively simple and non-invasive (Berna et al., 2015). Breath sampling methods must, however, take into consideration the diffusion of volatile organic compounds (VOCs) from blood to alveolar air (where the alveolar breath is the ca. 350 mL end portion of a breath), which depend on polarity, Henry's law partition constants (air-water partition coefficients), solubility in fat, and volatility (Buszewski et al., 2008). Cao and Duran (2006) proposed the calculation of partition coefficients (and factors affecting these coefficients which should be taken into account) for all compounds measured in breath. In addition, care should be taken to maximize sample quality by minimizing interferences from exogenous environmental VOCs by minimizing contamination from surroundings, nasal passages and the oral cavity itself (Pieil et al., 2013). Although the above is indeed recommended, the authors also propose that the sampling process should also aim to be as chemically unselective as possible.

FIGURE 7.1   Sample vials of breath condensate ready for testing for *H. pylori* infection.

Of note, it is important that the biological and analytical precision (or repeatability) are first determined and subsequently reported. This is typically done through reporting the intra- and inter-coefficients of variability (CV), which account for biological and analytical variations within the sampled cohort. In larger studies with many samples to be tested, it is necessary that samples be analyzed in batches, each with its own independent set of calibrators for the standard curve, and a series of breath controls with known concentrations of standards and surrogates. Most studies measure each sample in duplicate or triplicate for each analyte, with some researchers extending beyond this and analyzing each sample up to 10 times (more in some cases) (Ishikawa et al., 2016). Such a protocol ensures that both sample variation and analytical drift are captured and corrected; ensuring breath samples are scientifically valid.

However, a major pitfall in metabolomics-based research, as is the case in biological research in general, is the inability to reproduce results of published studies. This was discussed in detail in two well-publicized works by Ionnidis (2005) and Buttn et al. (2013), and it concluded that poor reproducibility occurred when sample power was low (i.e., experiments that are designed with a small sample size) and an observed smaller effect size between the compared groups were overvalued (i.e., the statistical significance that defines the difference between groups is close to the 0.05 p-value cut-off). Thus, in order to obtain statistically significant results that are reproducible, studies should ensure that a sample size is sufficiently large to reduce the variation between subjects within a group and to focus on findings that statistically show a substantial difference between two or more compared groups.

7.3   SAMPLING TECHNIQUES

In sampling breath, it is important to note that exhaled breath is a biological medium that potentially contains molecular information relating to health. For this to be true, the relevant metabolites need to freely transfer from the blood stream into the air within the gas-exchanging region of the lungs via the alveolar–capillary barrier (Daniel and Thangavel, 2016). Breath sampling methods range from initial experiments conducted by breathing directly into an analysis platform (Fantuzzi et al., 2001), to more recent collection methods using Tedlar® bags (Figure 7.2), aluminized Mylar bags, Tenax® cartridges, Bio-VOC™ bottles among others. Table 7.1 lists a range of commercial and non-commercial breath sampling devices. Furthermore, the following sections detail some of the common sampling approaches used for breath that are frequently used in metabolomics-based research focused on biomarker discovery for disease diagnostics. For example, the work from Pieil et al. (2013) used a device called the '*single breath canister*,' which comprised of a stainless steel canister with room for 1.0–1.5 L of alveolar breath. The device made use of a Teflon tube as a mouth piece, with isoprene and acetone being used as markers to assess the quality of exhaled breath. Earlier work by Dyne et al. (1997) had proposed a device which was portable, robust relied on the last part of the breath being held within a tube and sampled.

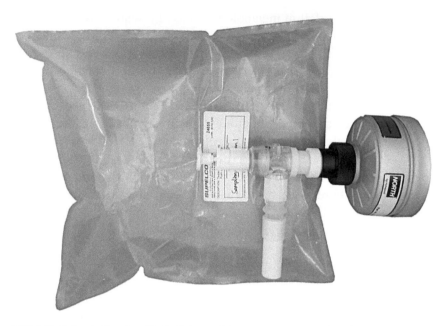

FIGURE 7.2 A Supelco Tedlar® bag with a two-way nonrebreathing valve (Hans Rudolph, Inc.) and ambient air filter for breath collection.

As the uptake of breath based research studies increased, so did the availability of commercial sampling products. For example, the 'Bio-VOC™ sampler' produced by Markes International, was based on pioneering work carried out at the UK Health & Safety Laboratory (Kwak et al., 2014). In this sampler, only the final part of the breath (ca. 150 mL) is retained and is assumed to come entirely from the alveolar portion of the lungs. The sample can then be pushed onto a sorbent tube (which is then capped) or Tedlar bag® through a connection to the opened end of the cylinder. This is essentially a Haldane-Priestley method. An alternative approach to alveolar air collection is to gate breath sampling based on measured expired carbon dioxide or pressure of sampled breath. An example of such a device is shown in Figure 7.3, in this system VOC-scrubbed air is supplied by a continuous positive airways pressure (CPAP) device and alveolar breath is selectively sampled on to sorbent tubes based breath profiles measured at the patient interface such that variable breathing patterns can be accommodated (Basanta et al., 2012).

TABLE 7.1 Commercial and Non-Commercial Breath Sample Collection and Pre-Concentration Devices for Breathomics-Based Research

Collection Device	Initial	Modified	Late/end	Mixed	Characteristics	References
ALTEF polypropylene bags.	◉	◎	◎	◉	Excellent low-cost alternative to Tedlar bags for collection of most VOCs.	(Harshman et al., 2015)
Aluminium gas bags.	◎	◎	◎	◎	Sampling and storing low MW compounds not stable in other bags.	(Berna et al., 2015)
BCA (breath collection apparatus).			◉	◎	Enable large volumes of breath to be captured on sorbent tubes.	(Dyne et al., 1997; Phillips et al., 1999)
Bio-VOC sampler.		◉	◎	◉	Simple-to-use device designed for sampling (VOCs).	(Kwak et al., 2010; Phillips et al., 2014)
FlexFilm bags.	◎	◎	◉	◎	Excellent low-cost alternative to Tedlar bags for collection of most VOCs.	(Bigazzi et al., 2016)
FEP bags (fluropolymer).	◎	◎	◎	◎	Alternative to Tedlar bags which are resistant to corrosive compounds.	(Schatz et al., 2016)
Gas sampling bulb.	◉			◉	Ideal for trapping and decanting an aliquot sample.	(Prado et al., 2003; Scott-Thomas et al., 2013)
Gas tight syringe (with glass vial and/or PP tubing).		◎	◉	◎	Ideal for trapping and decanting a small aliquot of sample.	(Buszewski et al., 2009; King et al., 2010)
Haldane-Priestly tube.		◉	◎		Alveolar gas obtained from the end of a maximal expiration into a Haldane tube.	(Beale et al., 2017a; Silva and Beyette, 2014)
Mylar bags.	◎	◎	◎	◎	Alternative to Tedlar bags for collection of most VOCs.	(Gruber et al., 2014)

*Expiratory Breath Type

TABLE 7.1 *(Continued)*

Collection Device	Expiratory Breath Type*				Characteristics	References
	Initial	Modified	Late/end	Mixed		
QuinTron sampler.		⊙		⊙	Commercially available, disposable, and suitable for home sampling.	(Beale et al., 2017a)
ReCIVA sampler.		⊙			Commercially available, repeatable and reproducible, storage on sorbent tubes.	(Beale et al., 2017a)
RTube sampler.		⊙	⊙	⊙	Designed for ease of use by unsupervised patient. Commercially available.	(Martin et al., 2010)
Smart Bag PA.	◎	⊙	◎	◎	Polymer film bag that delivers superior resistance to solvents/heat/adsorption.	(Ueta et al., 2014)
Supel foil bags.	◎	⊙	◎	◎	Alternative to Tedlar bags which are resistant to corrosive compounds.	(Scott-Thomas et al., 2013)
Tedlar bags.	⊙	⊙	⊙	⊙	Strong, durable, and chemically inert to a wide range of compounds.	(Amal et al., 2016; Baranska et al., 2016)
Tenax cartridges (GR/TA).	⊙	⊙	⊙	⊙	Specifically designed for the trapping of volatiles and semi-volatiles.	(Bigazzi et al., 2016; Ibrahim et al., 2013)
Carbotrap cartridges.	◎	⊙	◎	◎	Trapping a wide range of airborne organic compounds.	(Bigazzi et al., 2016)
Glass beads.	◎	◎	◎	◎	Used for trapping of large molecular weight compounds.	(Minh et al., 2011)

Note: *Modified expiratory breath is a type of alveolar breath sample. ⊙ Denotes that the device/technique has been successfully used in breathomics-based research published in the available scientific literature. ◎ Denotes the device/technique could be used/applied to sample the marked breath type.

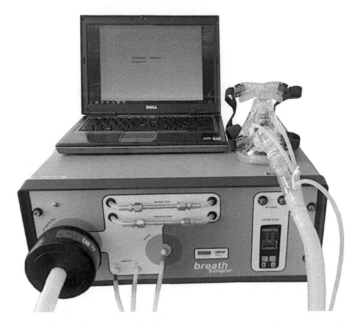

FIGURE 7.3 (See color insert.) An alveolar breath sampling system based on a modified CPAP device.

More recently, a 'breathe free' consortium (www.breathe-free.org) has been launched to establish a network of researchers interested in breath analysis, as well as address issues associated with data collation. In addition, the 'breathe free' consortium notes that the 'ReCIVA' (Respiration Collector for In Vitro Analysis) breath sampler has been developed (shown in Figure 7.4) and is currently in use in seven hospitals worldwide for the detection of lung cancer. Some of the advantages associated with the ReCIVA sampler include: flexibility that allows a selection of different breath volumes and fractions (alveolar, bronchial) to be collected; reproducible and repeatable sampling, and potential to store sorbent tubes for later analysis. However, it should be noted that further validation and head-to-head comparisons of the ReCIVA sampler with other devices are required before clinical applications can be developed.

A range of pre-concentration techniques consisting of stainless-steel/glass tubes (containing adsorbents used for various samples) is available for saliva/breath sampling. An adsorbent that is appropriate for all VOCs must be chosen to pack such tubes, and this is not always an easy task. There are four different main types of adsorbents: porous organic polymers, activated charcoal, carbon molecular sieves and graphitized carbon blacks (which can be porous or non-porous). Tenax (2,6-diphenyl-p-phenylene oxide) is arguably the most commonly used adsorbent due to its hydrophobicity, thermal stability and its ability to adsorb a wide range of VOCs.

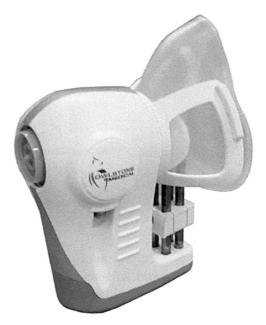

FIGURE 7.4   The ReCIVA breath sampler from Owlstone Medical.

Some researchers collect exhaled breath condensate (EBC), which is a biofluid obtained non-invasively after collecting and cooling the exhaled air (Baraldi et al., 2009; Carraro et al., 2007; Ibrahim et al., 2013). Typically, the condensate is collected via a sampling device fitted with a condenser and a saliva trap. A major advantage of analyzing EBC is that it captures both volatile and non-volatile metabolites (Nobakht et al., 2015). The EBC is collected over a period of 15 or more minutes and its composition is believed to reflect that of the fluid lining the airways (Carraro et al., 2007). Exhaled breath vapor/condensate (EBV/EBC) collection, has been described in a widely cited research paper by Martin et al. (2010). The study involved the use of a solid phase microextraction (SPME) fibre fitted inside the commercial breath collection device, namely the RTube™. The SPME adsorbed sample was then desorbed to a gas chromatography mass spectrometry (GC-MS) assembly for analysis. The test indicated a presence of limonene and related metabolites such as pinene, myrcene and terpinols from breath samples of individuals who had consumed lemonade. The study also showed a great potential for detecting compounds more relevant to medical diagnosis.

A recent review also indicated the importance of EBV/EBC in conjugation with techniques such as GC-MS, nuclear magnetic resonance (NMR) spectroscopy and sensor systems for analyzing a real time '*breathprint*' for diagnosing respiratory issues such as asthma, lung cancer, cystic fibrosis and chronic obstructive pulmonary disease (COPD) among others (Santini et al., 2016). Similar work described the use of breath VOCs as markers for irritable bowel syndrome (IBS). The experiment in question used Tedlar Bag® bags to collect the breath samples of patients (Baranska et al., 2016). A similar principle was also successfully utilized to detect and determine signature metabolite markers of *Plasmodium falciparum* infection (malaria) (Berna et al., 2015) and measure the direct relationship between the creatinine concentration and chronic kidney disease (CKD) (Qian et al., 2016). Overall, a number of studies have indicated that breath tests are the earliest and least invasive metabolomic profiling sampling methods. In the case of communicable diseases, further development of this technology may aid in preventing disease spread by enabling faster and earlier diagnosis.

7.4   INSTRUMENTATION AND ANALYTICAL TECHNIQUES

A major challenge in metabolomics is to address the extremely diverse and complex nature of the compounds being analyzed. The term 'metabolites' cover a huge range of biomolecules exhibiting a wide range of concentrations and physicochemical characteristics such as mass and polarity. More than 3,000 compounds have been reportedly detected in breath (although an '*average*' sample contains only around 200 compounds) (Phillips et al., 1999).

Unlike blood, urine or other biological fluids - breath (while also easy to sample) - is a particularly challenging sample matrix due to the low concentrations of metabolites present and the fact that many common analytical methods are not designed for gas analysis. Although breath condensate is mostly water, the volatile compounds in the gas phase are often the most diagnostically useful.

A variety of methods have been utilized to study breath ranging from the more commonly applied metabolomics techniques such as NMR, GC-MS and LC-MS through to more specialized technologies such as electronic noses. A selection of the more common approaches will be discussed here.

7.4.1   NUCLEAR MAGNETIC RESONANCE (NMR) SPECTROSCOPY

NMR spectroscopy is a powerful and well-used tool for molecular identification and structure analysis. It has been the mainstay of metabolomics/ metabonomics since the terms were first defined in the late 1990s (Nicholson et al., 1999; Oliver et al., 1998; Tweeddale et al., 1998). It is popular since it provides a relatively simple method for measuring the ensemble of metabolites present in a solution with minimal sample preparation and is relatively inexpensive on a per sample basis. For the interested reader, a more complete description of NMR including the classical and quantum mechanics involved can be found elsewhere (Hanson, 2008; Keeler, 2005).

NMR is primarily designed to work with samples in solution or solid state. This does not pose a problem for many biological fluid samples but it does for the analysis of breath samples, especially those containing only trace amounts of analytes. NMR poses several difficulties in terms of both sample handling and in low signal-to-noise ratio of the NMR signals obtained from gases at atmospheric pressure. NMR-based metabolic analysis of breath condensate has however, been carried out and a good review of the use of NMR-based metabolomics to explore airway diseases is given by (Sofia et al., 2011). In such cases, breath condensate, which contains predominantly water vapor but also volatile and non-volatile substances from the lower airways, is collected through spirometry or a condenser. Sample processing is minimal and usually only requires the addition to the solution of ~70 μL of deuterium oxide (D_2O or ^2H_2O) to provide a frequency lock for the spectrometer, a standard reference compound such as sodium-3-trimethylsilyl (2,2,3,3-2H4) propionate (TSP, 1 mM), which has a chemical shift of 0 ppm, and sodium azide (3 mM) to kill any microorganisms present in the sample. Although there is a large peak in the resulting spectra from the water in the sample, this can be suppressed via several common NMR pulse sequences. Statistical analysis via techniques such as principle component analysis (PCA) or partial least squares (PLS) can then be used to classify the data and potentially provide diagnostic information.

It should be noted, however, that there has been some discussion of whether NMR-based metabolomic analysis of EBC is clinically useful or not (Izquierdo-García et al., 2011; Motta et al., 2012). The concern has been that the cleaning protocols used on the reusable condenser parts to clean the system between patients could produce an artificial metabolic fingerprint not related to the endogenous metabolic pathways under study. It would appear, however, that although collection devices are an important source of variability of breath condensate analysis (Koczulla et al., 2009; Rosias et al., 2008), suitable cleaning procedures and data quality control procedures enable NMR analysis of breath condensate to be useful for metabolomic analyses. Lack of full standardization of collection methods and analysis techniques does, however, still hamper the introduction of such methods to routine clinical practice (van Mastrigt et al., 2015).

7.4.2   GAS AND LIQUID CHROMATOGRAPHY MASS SPECTROMETRY (GC-MS AND LC-MS)

GC-MS is one of the most widely used and powerful methods within the analytical sciences. It detects and quantifies gases from 100 parts per million (ppm) to 1 part per billion (ppb) or less, and has been a mainstay of metabolomics research for many years, where it is often seen as the 'gold standard' (Jones and Cheung, 2007). Some of the first comprehensive breath studies were performed using GC in 1971 by Pauling et al., who collected the gas from multiple breath exhalations in a cold tube then heated the tube and used GC to analyze the released gases (Pauling et al., 1971). Pauling found 250 substances in a sample of breath, and about 280 substances in a sample of urine vapor. It was thought that the new technique would be useful for medical purposes but at the time analyzing the data was too difficult and time-consuming, therefore the technique did not become adopted into clinical practice. Metabolomics, with its focus on combining analytical techniques, advanced statistical analysis and biological interpretation, may change this in the future.

As the name suggests, samples for GC-MS must either be in the gaseous phase or, more likely, be transferred into the gaseous phase by heating. The sample is then injected into the chromatograph where an inert carrier gas (usually helium, but increasingly hydrogen) is used to transport it through a packed or open, tubular (capillary) column. The column is typically coiled and very thin (0.25 mm internal diameter) allowing even those tens of meters in length to be housed within a relatively small temperature controlled oven. The exact length of the column depends on the type and speed of the desired analysis, but for metabolomic studies, longer columns (~30 meters) are generally used as these provide better chromatographic resolution and ensure maximum separation of the analytes in complex samples. Separation of compounds occurs due to differing rates of partitioning of the components of the sample between the internal lining of the column (stationary phase) and the carrier gas (mobile phase). This means each compound exits the column at a different time (known as the retention time). The mass spectrometer (MS) can then be used to detect eluting compounds, traditionally using electron impact (EI) ionization to ionize the compound, and then to measure the mass to charge ratio of each ion and generate a unique mass spectrum for the compound. SPME can also be used to absorb and pre-concentrate volatile compounds in breath prior to analysis.

LC-MS operates on a similar principle to GC-MS except that the chromatography stage utilizes a liquid mobile phase rather than a gas phase. This eliminates the need for metabolite volatility so there is no requirement for sample derivatization (see below), meaning that a much wider range of analytes can potentially be measured, the overall analysis time per sample is often much shorter than for GC-MS and a greater number of ionization mechanisms are possible. The use of liquid phase solvents also allows for a greater range of separation mechanisms than with non-reactive gases such as helium in GC. However, LC-MS can produce more variable data than GC-MS, due to larger retention time drifts.

Both GC-MS and LC-MS offer very high chromatographic resolution and good reproducibility of results (Buszewski et al., 2009; Spinhirne et al., 2004). An example GC chromatogram is shown in Figure 7.5. The techniques offer increased sensitivity compared with NMR whilst extensive and easily searchable libraries of molecular fragmentation patterns facilitate metabolite identification (Sweetlove et al., 2003). An added bonus for breath analysis via GC is that samples are already in the gas phase and thus do not require chemical derivatization prior to analysis (Fowler et al., 2015). For breath condensate samples, standard chemical derivatization is still needed to make the samples volatile and thermally stable enough for GC analysis and this significantly increases the sample preparation time.

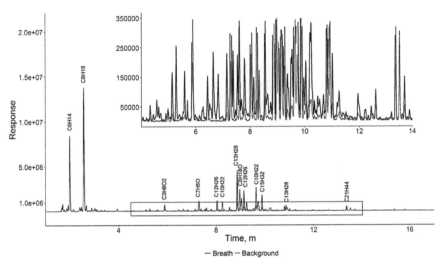

FIGURE 7.5   An example GC chromatogram for an extracted single ion (m/z = 57) for breath (black) and ambient air background (grey).

7.4.3 CAPILLARY ELECTROPHORESIS

Electrophoresis is defined as the migration of ions under the influence of an electric field, and gel electrophoresis is a commonly used method to separate proteins by size and charge (Kubáň et al., 2012). The development of capillary electrophoresis (CE) was enabled by advances in GC column technology, but instead of using pressurized gas or liquid as the mobile phase, CE uses high voltages to generate an electrophoretic flow of ionic species within a narrow-bore (20–200 μm i.d.) capillary. CE can be linked to either an ultraviolet–visible spectroscopy (UV-vis) or MS detector, and the resulting data look very similar to LC chromatograms; although in CE the data are referred to as an electropherogram rather than a chromatogram and retention time is referred to as migration time.

An often-overlooked advantage of CE is that its sample volumes are some of the smallest of any modern separation method due to the small diameter of the typical CE capillary. Sample volumes for LC are ~20 μL, those of capillary GC are ~0.5 to 1 μL but CE sample volumes can be as low as 100, 10, and even 1 nL or less. CE can also perform efficient separations of both large and small molecules making it a very versatile method.

The disadvantages of CE are that is it can be difficult to set up, and narrow capillary blockages often occur with buffer and salts. Its popularity has thus lagged behind other separation methods but a recent study using CE with conductometric detection used small volumes (100–200 µL) of exhaled breath to simultaneous determine inorganic cations and anions, and organic anions, such as chloride, nitrate, sulphate, lactate and potassium in patients with chronic obstructive pulmonary disease (COPD) (Kubáň et al., 2012).

7.4.4   NMR-GC HYPHENATION

NMR allows constitutional and configurational isomers to be distinguished but, while semi-preparative GC is an excellent technique for isolating small amounts of volatile compounds from mixtures such as breath, the recording of the NMR spectra of gases leaving a gas chromatograph, which contain only trace amounts of analytes imposes some sensitivity problems. Online coupling of GC with NMR has nevertheless been described for the analysis of volatile stereoisomers (Kühnle et al., 2008; Kühnle et al., 2010), caffeine (Kim et al., 2013), and menthol and menthone from peppermint oil (Park et al., 2012). Gas phase NMR is thus possible, although far too complex for regular use but further developments in this field are still highly desirable for breath analysis. Hyphenation of LC with NMR is also possible but is a more specialized technique, mainly used to perform isolation and structural determination in natural product chemistry (Dias and Urban, 2008; Urban and Dias, 2013) rather than for metabolomics studies.

7.4.5   DIRECT INJECTION MASS SPECTROMETRY

There are various forms of direct injection MS that have been used for the analysis of trace amounts of VOCs such as acetone, acetaldehyde, methanol, ethanol, benzene, toluene, xylene and inorganic gases in air and breath (Lamote et al., 2014). Such methods include secondary electrospray ionization-mass spectrometry (SESI-MS) (Zhu et al., 2013a), proton-transfer-reaction mass spectrometry (PTR-MS) (Zhan et al., 2013), and selected ion flow tube mass spectrometry (SIFT-MS) (Kumar et al., 2013).

SESI-MS is not widely used but both PTR-MS and SIFT-MS are popular methods of analysis in a range of areas including environmental, food and health sciences (Kumar et al., 2013; Zhan et al., 2013). In PTR-MS, a hollow cathode ion source produces H_3O^+ ions from high purity (>99%) distilled water. The reagent ions enter a drift tube where the trace compounds are ionized via proton transfer before analysis with either a quadrupole or a high-resolution time-of-flight mass analyzer. A disadvantage of PTR-MS is that is can only work for target molecules with a proton affinity higher than that of water. SIFT-MS is similar to PTR-MS but uses a greater number of precursor ions for chemical ionization (H_3O^+, NO^+, or O_2^+) and thus works with a wider range of exhaled metabolites. Both systems allow real-time, quantitative analysis and eliminate the need for sample preparation, pre-concentration and chromatography or other forms of separation. They do not, however, detect as many compounds as other forms of analysis such as GC and LC.

7.4.6 ION MOBILITY SPECTROMETRY

Ion mobility spectrometry (IMS) is an analytical technique used to separate and identify ionized molecules in the gas phase based on their mobility in a carrier buffer gas and, when coupled with MS (e.g., IMS-MS), enables additional separation of ions by their mass-to-charge (m/z) ratio (Lanucara et al., 2014). Furthermore, there are few variants of IMS that enable the use of an external electric field at ambient pressure and temperatures in order to separate different ions formed from the target analytes; these include differential mobility spectrometers (DMS) and high-field asymmetric waveform ion mobility spectrometers (FAIMS) (Pereira et al., 2015). Generally, IMS is commonly used to detect explosives, chemical warfare agents or illegal drugs but more recently has been used to analyze VOCs in breath (Halbfeld et al., 2014). Typically, VOC detection limits using IMS are of the magnitude of pg/L to ng/L-range, and when IMS is coupled with a multi-capillary column as a breath-sampling device, the analysis can be performed in under 8 min and at the site of sampling (Handa et al., 2014).

7.4.7 ULTRAVIOLET(UV)ANDINFRA-RED(IR)SPECTROSCOPY

Spectroscopy is based on the measurement of absorption of electro-magnetic radiation by a compound, or compounds of interest. Spectral fingerprints of compounds in breath span the UV to the mid-IR spectral regions. Typically, exhaled breath samples require pre-concentration prior to analysis via SPME, a suitable absorbent material or by direct cryofocusing (Miekisch and Schubert, 2006). Compounds present in exhaled breath that are IR or UV active such as ammonia, carbon monoxide, carbon dioxide, methane, and ethane absorb light at wavelengths characteristic of the bonds present in the molecule and these absorption bands can be used to identify specific molecular components and/or to allow identification of a compound via reference library matching. Stable isotopomers of IR active molecules can also be accurately detected, making it possible to follow specific metabolic processes. While infrared spectroscopy data are not as detailed as those from NMR or MS-based methods, the technique has the advantages that it is quick, simple, non-destructive and does not require extensive sample preparation; near real-time data can be obtained and the instruments are much lower in cost that NMR or MS instruments. The disadvantages are that spectroscopy is not as sensitive or selective as MS, with detection limits in the ppm to ppb range and the technique is also limited in the number of chemical species it can distinguish.

Although specialized methods such as cavity ring-down spectroscopy (in which light is trapped for several microseconds between two highly reflective mirrors) can increase the sensitivity of spectroscopic techniques, these methods are not widely used and still do not detect as wide range of compounds as MS. Breath analysis using laser spectroscopic techniques is only a very recent advancement and, to a great extent, relies on recent development of diode lasers capable of covering a wide spectral range of molecular fingerprints (Wang and Sahay, 2009). Comprehensive reviews of breath analysis using laser spectroscopic techniques are available in Wang and Sahay (2009), and Chow et al. (2012). Raman spectroscopy has also shown promise for the analysis of breath samples but still requires further development before it is commonly applied (Carlsen et al., 1995; Hanf et al., 2014).

7.4.8 INDUCTIVELY COUPLED PLASMA MASS SPECTROMETRY (ICP-MS)

Inductively coupled plasma-mass spectrometry (ICP-MS) has been one of the recent but rapidly growing techniques used in quantitative metabolomics-based analyses. Similar to the previously discussed techniques, ICP-MS is considered a highly sensitive and selective method. Although it is predominately used as a standalone analytical tool for analyzing metallic entities in geological and metallurgical samples (Beauchemin, 2004), it has recently been applied to the biological samples (Wang et al., 2017). The technique uses an inductively coupled argon plasma-based ionization source, which is connected to a mass spectrometer. The sample passes through a nebulizer and ICP source, causing a rapid ionization for further analysis. Due to the inductive plasma nature of ionization, the technique is also able to provide a clear difference between various isotopes, at great sensitivities, at parts per trillion (ppt) concentrations (Sakata et al., 2001).

Due to the very broad nature of analytical capabilities, the sensitivity of ICP-MS-based metabolomics suffers to an extent when applied to biological samples. These complications, however, can be resolved by a number of pre-treatment techniques, which are used to remove the background matrix and increase the relative abundance of targeted entities. These methods include, but are not limited to, SPME, liquid phase extraction, and single drop microextraction. Furthermore, ICP-MS coupled with other gold standard techniques such as GC-MS, LC-MS and CE enables another dimension for characterization of biological samples such as urine, saliva (Huang et al., 2014) and exhaled breath (Wilson, 2014). The reader is referred to a recent review by Wang et al. (2017) for further details.

7.4.9 ELECTRONIC NOSES

Electronic noses (E-noses) are artificial sensor systems, usually consisting of a range of sensors for various chemicals of interest. E-noses are able to detect ('*smell*') patterns of VOCs in breath and then use algorithms for classification of the '*breathprint*' and comparison with previously recorded samples from known sources. Such methods can be paired with, and add value to existing diagnostic tests, such as routine spirometry (Vries et al., 2015). Although a relatively new technique, E-noses have been used to discriminate between patients with respiratory disease, including asthma, COPD and lung cancer, and healthy control subjects, and also among patients with different respiratory diseases and with airway inflammation activity (Montuschi et al., 2013).

7.5   BREATHOMICS APPLICATIONS

While the total number of breath related metabolomics-based studies are limited when compared to all metabolomics-based studies, there is still a wide breadth of breathomics-based research related to disease diagnostics and characterization. The following section provides a few applications and the potential of this technology for the development of rapid diagnostic methods and personalized medicine (Bujak et al., 2015).

7.5.1   *INFECTIOUS DISEASES*

Owing to the complexity of the VOCs expelled through breath, breathomics-based research can inform on the different stages of infection (i.e., early-to-intermediate), allowing more time for diagnosis and treatment of infections. For example, there has been recent progress in the detection of infectious diseases with higher social impacts such as tuberculosis, especially in developing countries. Maiga et al. (2014) utilized the conversion of isotopic ^{13}CO to $^{13}CO_2$ by the pathogen *Mycobacterium bovis* in rabbit populations. The testing indicated that detection of CO_2 generated through the activity of mycobacterial carbon monoxide dehydrogenase (CODH) had the potential to provide rapid and non-invasive diagnosis of tuberculosis (Maiga et al. 2014). While this study demonstrated the preclinical breath analysis of *Mycobacterium bovis* in rabbit populations, if developed further it could enable for cheaper and faster point-of-care diagnosis of tuberculosis within a clinical setting, which would augment and improve current pathology tests. Zhu et al. (2013b) extended breath single biomarker analysis further and used the entire *'breathprint'* to diagnose acute *Pseudomonas aeruginosa* and *Staphylococcus aureus* lung infections in a mouse model. Such an approach that uses the entire *'breathprint'* rather than single biomarkers enables volatile metabolites to be characterized and monitored during the course of infection, and potentially identify a suite of breath biomarkers that can be used in the diagnoses of pathogens at any point during the infection (Bean et al., 2015; Zhu et al., 2013a, 2013b).

A recent study by Berna et al. (2015) investigated the relationship between *Plasmodium falciparum* (the parasite that causes malaria) and the exhaled breath of individuals infected with the parasite within a controlled clinical study. The current approach for malarial diagnosis involves an assessment of the patient's clinical presentation and an analysis of a sample of blood via microscopy; which is both time consuming and expensive. In this study, breath samples were collected using sorbent tubes and analyzed via GC-MS to detect specific malaria-associated VOCs. Nine compounds were identified and their concentrations varied significantly over the course of the infection. The identified metabolites comprised: carbon dioxide, isoprene, acetone, benzene, cyclohexanone, allyl methyl sulphide, 1-methylthio-propane, (Z)-1-methylthio-1-propene, and (E)-1-methylthio-1-propene (Berna et al., 2015). If the work could be extended and the biomarkers shown to be unambiguously related to malaria infection, the method could be used to screen a patient's breath for the presence of malaria, which would be a rapid and cost effective tool in developing countries where access to medical treatment is both geographically difficult and cost limited (Berna et al., 2016).

7.5.2 INFLUENZA INFECTION AND POSSIBLE ANTIVIRAL APPLICATIONS

Influenza is a highly contagious respiratory disease that causes high global morbidity and mortality (Lowen et al., 2006). Understanding the pathogenesis of influenza virus is critical for effective disease control in a pandemic scenario, enables the screening of the emergence of new strains with pandemic potential, and facilitates the development of vaccines and antiviral drugs. The application of metabolomics to study the dynamics of influenza infection with host metabolism is in its infancy, with the majority of work to date being done using viruses grown in cell cultures (Chen et al., 2014; Chung et al., 2016; Fu et al., 2016; Lin et al., 2010; Rabinowitz et al., 2011; Ritter et al., 2010). However, Aksenov et al. (2014) examined VOCs produced directly at the cellular level from B lymphoblastoid cells upon infection with three influenza A virus subtypes [H9N2 (avian), H6N2 (avian), and H1N1 (human)] (Aksenov et al., 2014) via headspace GC-MS. The patterns of VOCs produced in response to infection were unique for each subtype, and the metabolic flux of the VOC released post infection were different. The emitted VOCs included esters and other oxygenated compounds, which was attributed to the increased oxidative stress resulting from the viral infection. It was concluded that elucidating such VOC signatures from the host cell's response to infection may yield non-invasive diagnostics of influenza and other viral infections

(Aksenov et al., 2014); this approach has been applied for the determination of *Plasmodium falciparum* (malaria) in humans (Berna et al., 2016) and *Acinetobacter baumannii* colonization in the lower respiratory tract (which results in ventilator-associated pneumonia in patients admitted to hospital) (Jianping et al., 2016).

Progressing from *in vitro* studies, animal models of influenza infection such as the mouse, have been used to study metabolic changes associated with influenza infection at an organism level (Chandler et al., 2016; Cui et al., 2016). In mice, influenza infection was associated with both systemic (serum) and localized (lung and bronchoalveolar lavage) changes of more than 100 metabolites that were associated with the pulmonary surfactant system, suggesting viral-induced lung injury (Cui et al., 2016). A metabolome-wide association study with cytokines using high-resolution metabolomics in the influenza-infected mouse lung also revealed high correlation of 396 metabolites with proinflammatory cytokines, such as IFNγ, IL-1β, TNF-α and anti-inflammatory cytokines, such as IL-10 (Chandler et al., 2016). In humans, metabolomics has been applied as a diagnostic tool for acute respiratory distress syndrome caused by influenza A(H1N1) infection via NMR analysis of collected blood serum (Jose et al., 2012). Using GC-MS analysis of the breath collected from individuals that were vaccinated with live attenuated influenza, Philips et al. (2010) reported a distinct VOC signature in the vaccinated group showing the potential to use breathomics to diagnose influenza infection (Phillips et al., 2010). More studies are needed using breathomics-based approaches within a clinical setting for influenza studies, in particular investigating the efficacy of antiviral treatments and the measurable effects on the breathprint.

7.5.3 CANCER AND DEGENERATIVE DISEASES

Breath testing techniques such as electronic nose devices have been used to detect and differentiate the patients suffering from Alzheimer's and Parkinson's diseases (Bach et al., 2015). The study used an E-nose (for VOC sampling) in conjugation with ion mobility spectrometry. The sample analysis was confirmed by use of an existing standard enzyme-linked immunosorbent assay (ELISA) technique for the detection of the possible causative agent, amyloid beta (Aβ). The predictability of the conducted tests to detect Alzheimer's and Parkinson's diseases was reported to be ca. 94%. Likewise, other chronic degenerative diseases such as liver cirrhosis can be detected at a comparatively early stage by VOC sample analysis (Pijls et al., 2016). The experiment involved the detection of signature metabolome profiles for chronic liver disease (CLD) and compensated cirrhosis (CC). GC-MS analysis was performed on exhaled breath collected in Tedlar® bags. Approximately 11 discriminatory metabolites were observed for CLD and CC. These metabolites comprised propanoic acid, octane, terpene, dimethyl sulfoxide (DMSO, increased), methyl butanal and hexadecanol (depleted), possibly caused by impaired cytochrome P450 detoxification mechanisms in the liver (Pijls et al., 2016).

Lung cancer has been one of the areas where the breath-based metabolomics has been extensively applied. Both COPD and lung cancer are classified as chronic respiratory disorders. However, biochemical tests such as immunoassays have not proven to be adequately efficient to selectively identify lung cancer from COPD. There have been multiple studies using metabolic analyses for detecting and characterizing markers for COPD and lung cancer. In one of these studies, hydrogen peroxide and 8-isoprostane were detected as the discriminatory factors between normal and COPD samples (Kostikas et al., 2003). Similarly, selective alkanes, aldehydes and other metabolites such as acetone have been reported. However, in almost all cases, the prediction in terms of analysis sensitivity and specificity of these tests has been limited (Horváth et al., 2009). Recent research conducted by Peralbo-Molina et al. (2016) involved the analysis of breath samples from smoking, non-smoking and past smoking patients in order to detect the metabolic signature within EBC for risk and cancer affected individuals. In addition to the previously detected alkanes (Horváth et al., 2009), a high-resolution GC-TOF(time of flight)-MS approach was able to identify monog lycerols and squalene as the prominent signature metabolites. Although the sensitivity and specificity of these tests were noticeably higher than the previously reported values, sensitivity was still below 85% – however, in the case of a distal airway metabolite analysis, specificity of about 90% was observed.

Due to the limitations of breath analysis, methods using breathomics in combination with cellular sampling/cell culture metabolomics will increase the sensitivity, specificity and, as a result, the predictability of cancer diagnostic metabolomics have been proposed. In a recent review, the small number of signature metabolites was identified as one of the major limitations of 'breath only' analyses. Kalluri et al. (2014) suggested that metabolic profiling of cell cultures and breath analysis be used in parallel to determine the signature metabolites attributed to lung cancer. The study indicates the origin of previously indicated signature metabolites from both healthy and affected individuals; the metabolites include aldehydes (formaldehyde, acetaldehyde), ketones (acetone) and various alkanes, which are formed during cellular oxidative stress, as observed during lung cancer. In such cases, cellular hypoxia has been observed, where sugars are metabolized in the absence of oxygen via glycolysis only and the Krebs cycle remains unutilized. Secretion of the above mentioned metabolites in addition to others such as excessive lactic acid (leading to cellular autophagy) can be used as more reliable indicators of lung cancer as compared to standalone breathomics. For a detailed overview of cancer signature metabolites, the reader is referred to a recent review by Patel and Ahmed (2015).

7.5.4 ASTHMA

Asthma is a widespread disease with estimates suggesting as many as 334 million people are affected globally. Asthma causes significant socioeconomic impacts with growing direct health-related costs of treating and managing affected people and indirect costs relating to lost productivity (absence from school and work) (Soriano et al., 2017). Current approaches used to diagnose and monitor asthma include a physical examination and a range of lung function tests, allergy testing, exhaled nitric oxide (ENO) tests, imaging, sputum eosinophils [count of white blood cells in sputum (saliva and mucus)], and provocative testing (exercise and cold-induced asthma). The time for diagnosis can be rather drawn out and, as such, metabolomics provides a unique opportunity where molecular determinants can be used to rapidly diagnose asthma and other respiratory illnesses (Dahlin et al., 2015; Luxon, 2014). To date, medical metabolomics studies have been limited in size and scope, with a large proportion of research focused on quantifying the small biochemicals in plasma and tissue samples in order to identify biomarkers that may serve as therapeutic targets (Comhair et al., 2015). In more recent times, the focus for metabolomics-based research has been on using breath samples, amongst other non-invasive sample matrices, to diagnose asthma from other respiratory illnesses amongst children (Adamko et al., 2012).

In a study by Carraro et al. (2007), NMR-based metabolomic analyses of EBC was used to discriminate between asthmatic and healthy children within a pediatric clinical setting (Baraldi et al., 2009). In the study, 25 children with asthma and 11 age-matched healthy control children were recruited. Each child provided a sample using an EBC condenser over a period of 15 min. The collected samples were then dried prior to reconstitution in D_2O for NMR analysis. In addition to the EBC sample, the children were tested using conventional asthma techniques such as ENO and measuring lung function via spirometry. Typically, the combination of ENO and lung function testing discriminates children with asthma and healthy children with a success rate of ca. 80%, whereas selected signals from NMR spectra offer a slightly better discrimination (ca. 86%). The selected NMR variables derive from the region of 3.2 to 3.4 ppm, indicative of oxidized compounds, and from the region of 1.7 to 2.2 ppm, indicative of acetylated compounds (Baraldi et al., 2009). In a similar study, 57 asthmatics and 22 healthy controls were recruited and EBC samples collected and analyzed by NMR. Thirteen spectral regions were identified as discriminating between asthmatics and the healthy control cohort. The normalized peak areas were used in multivariate logistic regression, and a model consisting of five regions predicted the asthmatics (82.3%) (Ibrahim et al., 2013). It is noted that the authors did not attempt to identify the compounds in the discriminating spectral regions. Motta et al. (2014) investigated two different condenser trap temperatures (−4°C and −37°C) and evaluated the effect using NMR (Motta et al., 2014). It was found that the two different temperatures resulted in no significant variation between the analyzed EBC when analyzed separately and as part of a blind analysis.

7.5.5 PULMONARY ARTERIAL HYPERTENSION (PAH)

PAH is a life-threatening condition in which patients present with a high blood pressure in the pulmonary arteries that go from the heart to the lungs (Guillevin et al., 2013). PAH result of the small arteries in the lungs becoming narrow or blocked, which in turn increases blood pressure in the lungs and puts stress on the heart as it works harder to pump blood to the lungs. Congestive heart failure, blood clots in the lungs, HIV, drug abuse, liver disease, autoimmune diseases, heart defects, lung diseases and sleep apnea have all been linked to PAH (Guillevin et al., 2013). Current approaches used to diagnose PAH include a physical examination and a range of specialist tests, such as: echocardiogram (ultrasound), computed tomography (CT) scan, ventilation-perfusion scan (to identify blood clots), electrocardiogram (ECG), chest X-ray and provocative testing (exercise).

Zhao et al. (2014) investigated the application of metabolomics as a diagnostic tool in human lung tissue samples collected from patients diagnosed with PAH and age-matched controls. Using a combination of LC-MS and GC-MS, it was identified that patients diagnosed with PAH showed unbiased metabolomic profiles of disrupted glycolysis, increased tricarboxylic acid (TCA) cycle, and fatty acid metabolites with altered oxidation pathways; indicating increased adenosine triphosphate (ATP) synthesis. It was concluded that these biomarkers could be used for the diagnosis of PAH, however, collecting lung tissue samples is considered invasive. An alternative approach would be to analyze the exhaled breath of patients diagnosed with PAH. Such an approach was undertaken by Cikach et al. (2014), where fasting state breath samples were collected and analyzed by SIFT-MS. It was found that the concentrations of the exhaled ammonia, 2-propanol, acetaldehyde, ammonia, ethanol, pentane, 1-decene, 1-octene, and 2-nonene were significantly different in patients with PAH compared to the control cohort (Cikach et al., 2014), with the concentration of compounds correlating with the severity of PAH. This suggests that differences in the breath metabolic profile can potentially be used to diagnose and classify the severity of PAH. Furthermore, with such observed differences, there is potential for the development of rapid diagnostic breath analyzers that could be used to monitor PAH progression. Such metabolomic approaches and tool development are also being applied to other respiratory diseases, such as acute respiratory distress syndrome (ARDS) (Bos et al., 2014; Stringer et al., 2016), COPD (Santini et al., 2016), lung cancer (Peralbo-Molina et al., 2016), or diseases with a clinically relevant respiratory component including cystic fibrosis (Montuschi et al., 2012; Muhlebach and Sha, 2015) and primary ciliary dyskinesia (Montuschi et al., 2014; Paris et al., 2015).

7.6   FUTURE DIRECTION AND CONCLUSIONS

In many fields of medicine, there is growing interest in characterizing diseases at the molecular level with a view to developing an individually tailored therapeutic approach. A comprehensive metabolome analysis, using a range of semi- or non-invasive sample methodologies such as breath would result in a greater depth of knowledge for specific diseases and disorders. Such an approach would enable the tracking of metabolic pathways, monitor the effectiveness of therapeutic treatments and interventions, and assess their impact on the entire biological system.

Through the application of breathomics-based techniques, it is envisaged that biomarkers will ultimately be identified in breath samples that are specific to a range of respiratory diseases and disorders, in addition to other systemic health conditions. Through such biomarker discovery and validation using traditional diagnostic methods, it is anticipated that a range of rapid diagnostic methods such as colorimetric assays and targeted metabolite breathalyzer technologies will be developed. Such biomarker discovery and assay/technologies can be used to develop rapid test kits for disease diagnoses that can be used in clinics to identify specific infections and diseases (such as malaria, PAH, cancer, etc.) and enable earlier medical intervention.

Furthermore, breath analysis is a tool which can potentially be used for human exposure assessment. Although efforts have been made to optimize breath analysis methods, there is still a need for more research demonstrating their suitability before these methods can be used routinely (validation studies). These studies should involve the standardization of collection methods and profiling via the various detection platforms available. Multiple efficient devices have also been developed which have shown potential. However, there are still issues involving leakage, adsorption and transfer processes. Lastly, the use of more sensitive and portable methods should allow for accurate identification and quantitation within a clinical environment, thus facilitating effective point-of-care testing.

ACKNOWLEDGMENTS

The authors would like to acknowledge and thank the Commonwealth Scientific and Industrial Research Organization (CSIRO), a federally funded research organization of the Australian Government, and the Melbourne WHO Collaborating Centre for Reference and Research on Influenza, which is supported by the Australian Government Department of Health, their support in preparing this book chapter. The authors would also like to acknowledge and thank Mr Jake Collie and Mr Max Willkinson for taking the photographs used in this chapter. No specific research grants are associated with the work presented.

KEYWORDS

- applications
- breathomics
- diagnosis
- disease
- instrumentation
- sampling techniques

REFERENCES

Adamko, D. J., Sykes, B. D., & Rowe, B. H., (2012). The metabolomics of asthma: novel diagnostic potential. *Chest, 141*(5), 1295–1302.

Aksenov, A. A., Sandrock, C. E., Zhao, W., Sankaran, S., Schivo, M., Harper, R., Cardona, C. J., Xing, Z., & Davis, C. E., (2014). Cellular scent of influenza virus infection. *Chembiochem, 15*(7), 1040–1048.

Amal, H., Leja, M., Funka, K., Lasina, I., Skapars, R., Sivins, A., Ancans, G., Kikuste, I., Vanags, A., Tolmanis, I., Kirsners, A., Kupcinskas, L., & Haick, H., (2016). Breath testing as potential colorectal cancer screening tool. *Int. J. Cancer., 138*(1), 229–236.

Bach, J. P., Gold, M., Mengel, D., Hattesohl, A., Lubbe, D., Schmid, S., Tackenberg, B., Rieke, J., Maddula, S., Baumbach, J. I., Nell, C., Boeselt, T., Michelis, J., Alferink, J., Heneka, M., Oertel, W., Jessen, F., Janciauskiene, S., Vogelmeier, C., Dodel, R., & Koczulla, A. R., (2015). Measuring compounds in exhaled air to detect Alzheimer's disease and Parkinson's disease. *PLoS One., 10*(7), e0132227.

Baraldi, E., Carraro, S., Giordano, G., Reniero, F., Perilongo, G., & Zacchello, F., (2009). Metabolomics: moving towards personalized medicine. *Ital. J. Pediatr., 35*(1), 1–4.

Baranska, A., Mujagic, Z., Smolinska, A., Dallinga, J., Jonkers, D., Tigchelaar, E., Dekens, J., Zhernakova, A., Ludwig, T., & Masclee, A., (2016). Volatile organic compounds in breath as markers for irritable bowel syndrome: a metabolomic approach. *Aliment. Pharmacol. Ther., 44*(1), 45–56.

Basanta, M., Ibrahim, B., Douce, D., Morris, M., Woodcock, A., & Fowler, S. J., (2012). Methodology validation, intra-subject reproducibility and stability of exhaled volatile organic compounds. *J. Breath Res., 6*(2), 026002.

Beale, D. J., Karpe, A. V., & Ahmed, W., (2016). Beyond metabolomics: A review of multi-omics-based approaches. In: *Microbial Metabolomics: Applications in Clinical, Environmental, and Industrial Microbiology*, Beale, D., Kouremenos, K., Palombo, E., eds. Springer International Publishing: Switzerland, pp. 295–319.

Beale, D., Jones, O., Karpe, A., Dayalan, S., Oh, D., Kouremenos, K., Ahmed, W., & Palombo, E., (2017a). A review of analytical techniques and their application in disease diagnosis in breathomics and salivaomics research. *Int. J. Mol. Sci., 18*(1), 24.

Beale, D., Karpe, A., Ahmed, W., Cook, S., Morrison, P., Staley, C., Sadowsky, M., & Palombo, E., (2017b). A community multi-omics approach towards the assessment of surface water quality in an urban river system. *Int. J. Environ. Res. Public Health, 14*(3), 303.

Bean, H. D., Jiménez-Díaz, J., Zhu, J., & Hill, J. E., (2015). Breathprints of model murine bacterial lung infections are linked with immune response. *Eur. Resp. J., 45*(1), 181–190.

Beauchemin, D., (2004). Inductively coupled plasma mass spectrometry. *Anal. Chem., 76*(12), 3395–3416.

Berna, A. Z., McCarthy, J. S., & Trowell, S. C., (2016). Malaria detection using breath biomarkers. *Med. J. Austral., 204*(2), 50.

Berna, A. Z., McCarthy, J. S., Wang, R. X., Saliba, K. J., Bravo, F. G., Cassells, J., Padovan, B., & Trowell, S. C., (2015). Analysis of breath specimens for biomarkers of *plasmodium falciparum* infection. *J. Infect. Dis., 212*(7), 1120–1128.

Bigazzi, A. Y., Figliozzi, M. A., Luo, W., & Pankow, J. F., (2016). Breath biomarkers to measure uptake of volatile organic compounds by bicyclists. *Environ. Sci. Technol., 50*(10), 5357–5363.

Bos, L. D., Weda, H., Wang, Y., Knobel, H. H., Nijsen, T. M., Vink, T. J., Zwinderman, A. H., Sterk, P. J., & Schultz, M. J., (2014). Exhaled breath metabolomics as a noninvasive diagnostic tool for acute respiratory distress syndrome. *Eur. Respir. J., 44*(1), 188–197.

Bujak, R., Struck-Lewicka, W., Markuszewski, M. J., & Kaliszan, R., (2015). Metabolomics for laboratory diagnostics. *J. Pharm. Biomed. Anal., 113*, 108–120.

Buszewski, B., Kesy, M., Ligor, T., & Amann, A., (2008). Human exhaled air analytics: Biomarkers of diseases. *Biomed. Chromatogr., 21*, 553–566.

Buszewski, B., Ulanowska, A., Ligor, T., Denderz, N., & Amann, A., (2009). Analysis of exhaled breath from smokers, passive smokers and non-smokers by solid-phase microextraction gas chromatography/mass spectrometry. *Biomed. Chromatogr., 23*(5), 551–556.

Button, K. S., Ioannidis, J. P. A., Mokrysz, C., Nosek, B. A., Flint, J., Robinson, E. S. J., & Munafo, M. R., (2013). Power failure: why small sample size undermines the reliability of neuroscience. *Nat. Rev. Neurosci., 14*(5), 365–376.

Cao, W. Q., & Duan, Y. X., (2006). Breath analysis: potential for clinical diagnosis and exposure assessment. *Clin. Chem., 52*(5), 800–811.

Carlsen, W. F., Simons, T. D., Pittaro, R. J., Perry, J., George, W. H. I. I., & Gray, D. F., (1995). Raman spectroscopy of airway gases. *US Patent 08, 226*, 245.

Carraro, S., Rezzi, S., Reniero, F., Héberger, K., Giordano, G., Zanconato, S., Guillou, C., & Baraldi, E., (2007). Metabolomics applied to exhaled breath condensate in childhood asthma. *Am. J. Resp. Crit. Care. Med., 175*(10), 986–990.

Chandler, J. D., Hu, X., Ko, E. J., Park, S., Lee, Y. T., Orr, M. L., Fernandes, J., Uppal, K., Kang, S. M., Jones, D. P., & Go, Y. M., (2016). Metabolic pathways of lung inflammation revealed by high-resolution metabolomics (HRM) of H1N1 influenza virus infection in mice. *Am. J. Physiol. Regul. Integr. Comp. Physiol., 311*(5), R906–R916.

Chen, L., Fan, J., Li, Y., Shi, X., Ju, D., Yan, Q., Yan, X., Han, L., & Zhu, H., (2014). Modified Jiu Wei Qiang Huo decoction improves dysfunctional metabolomics in influenza a pneumonia-infected mice. *Biomed. Chromatogr., 28*, 468–474.

Chow, K. K., Short, M., & Zeng, H., (2012). A comparison of spectroscopic techniques for human breath analysis. *Biomed. Spectrosc. Imaging., 1*(4), 339–353.

Chung, D. H., Golden, J. E., Adcock, R. S., Schroeder, C. E., Chu, Y. K., Sotsky, J. B., Cramer, D. E., Chilton, P. M., Song, C., Anantpadma, M., Davey, R. A., Prodhan, A. I., Yin, X., & Zhang, X., (2016). Discovery of a broad-spectrum antiviral compound that inhibits pyrimidine biosynthesis and establishes a Type 1 interferon-independent antiviral state. *Antimicrob. Agents Chemother.*, *60*(8), 4552–4562.

Cikach, F. S., Tonelli, A. R., Barnes, J., Paschke, K., Newman, J., Grove, D., Dababneh, L., Wang, S., & Dweik, R. A., (2014). Breath analysis in pulmonary arterial hypertension. *Chest*, *145*(3), 551–558.

Comhair, S. A., McDunn, J., Bennett, C., Fettig, J., Erzurum, S. C., & Kalhan, S. C., (2015). Metabolomic endotype of asthma. *J. Immunol.*, *195*(2), 643–650.

Cui, L., Zheng, D., Lee, Y. H., Chan, T. K., Kumar, Y., Ho, W. E., Chen, J. Z., Tannenbaum, S. R., & Ong, C. N., (2016). Metabolomics investigation reveals metabolite mediators associated with acute lung injury and repair in a murine model of influenza pneumonia. *Sci. Rep.*, *6*, 26076.

Dahlin, A., McGeachie, M. J., & Lasky-Su, J. A., (2015). Asthma metabolomics: The missing step for translating bench work into the clinic. *J. Pulm. Respir. Med.*, *5*(3), 267.

Daniel, D. A. P., & Thangavel, K., (2016). Breathomics for gastric cancer classification using back-propagation neural network. *J. Med. Signals Sens.*, *6*(3), 172–182.

Dias, D., & Urban, S., (2008). Phytochemical analysis of the Southern Australian marine alga, *Plocamium mertensii* using HPLC-NMR. *Phytochem. Anal.*, *19*(5), 453–470.

Dyne, D., Cocker, J., & Wilson, H., (1997). A novel device for capturing breath samples for solvent analysis. *Sci. Total. Environ. Health.*, *199*(1), 83–89.

Fantuzzi, G., Righi, E., Predieri, G., Ceppelli, G., Gobba, F., & Aggazzotti, G., (2001). Occupational exposure to trihalomethanes in indoor swimming pools. *Sci. Total. Environ. Health*, *264*(3), 257–265.

Fowler, S. J., Basanta-Sanchez, M., Xu, Y., Goodacre, R., & Dark, P. M., (2015). Surveillance for lower airway pathogens in mechanically ventilated patients by metabolomic analysis of exhaled breath: a case-control study. *Thorax*, *70*(4), 320–325.

Fu, Y., Gaelings, L., Soderholm, S., Belanov, S., Nandania, J., Nyman, T. A., Matikainen, S., Anders, S., Velagapudi, V., & Kainov, D. E., (2016). JNJ872 inhibits influenza A virus replication without altering cellular antiviral responses. *Antiviral Res.*, *133*, 23–31.

Goddard, A. F., & Logan, R. P. H., (1997). Urea breath tests for detecting *Helicobacter pylori*. *Aliment. Pharmacol. Ther.*, *11*(4), 641–649.

Gruber, M., Tisch, U., Jeries, R., Amal, H., Hakim, M., Ronen, O., Marshak, T., Zimmerman, D., Israel, O., Amiga, E., Doweck, I., & Haick, H., (2014). Analysis of exhaled breath for diagnosing head and neck squamous cell carcinoma: a feasibility study. *Br. J. Cancer.*, *111*(4), 790–798.

Guillevin, L., Armstrong, I., Aldrighetti, R., Howard, L. S., Ryftenius, H., Fischer, A., Lombardi, S., Studer, S., & Ferrari, P., (2013). Understanding the impact of pulmonary arterial hypertension on patients' and carers' lives. *Eur. Respir. Rev.*, *22*(130), 535–542.

Halbfeld, C., Ebert, B., & Blank, L., (2014). Multi-capillary column-ion mobility spectrometry of volatile metabolites emitted by *Saccharomyces Cerevisiae*. *Metabolites*, *4*(3), 751.

Handa, H., Usuba, A., Maddula, S., Baumbach, J. I., Mineshita, M., & Miyazawa, T., (2014). Exhaled breath analysis for lung cancer detection using ion mobility spectrometry. *PLoS One.*, *9*(12), e114555.

Hanf, S., Keiner, R., Yan, D., Popp, J., & Frosch, T., (2014). Fiber-enhanced Raman Multigas spectroscopy: A versatile tool for environmental gas sensing and breath analysis. *Anal. Chem.*, *86*(11), 5278–5285.

Hanson, L. G., (2008). Is quantum mechanics necessary for understanding magnetic resonance? *Concepts Magn. Reson.*, *32A*, 329–340.

Harshman, S. W., Geier, B. A., Fan, M., Rinehardt, S., Watts, B. S., Drummond, L. A., Preti, G., Phillips, J. B., Ott, D. K., & Grigsby, C. C., (2015). The identification of hypoxia biomarkers from exhaled breath under normobaric conditions. *J. Breath Res.*, *9*(4), 047103.

Horváth, I., Lázár, Z., Gyulai, N., Kollai, M., & Losonczy, G., (2009). Exhaled biomarkers in lung cancer. *Eur. Respir. J.*, *34*(1), 261–275.

Huang, Y., Zhu, M., Li, Z., Sa, R., Chu, Q., Zhang, Q., Zhang, H., Tang, W., Zhang, M., & Yin, H., (2014). Mass spectrometry-based metabolomic profiling identifies alterations in salivary redox status and fatty acid metabolism in response to inflammation and oxidative stress in periodontal disease. *Free Radical Bio. Med.*, *70*, 223–232.

Ibrahim, B., Marsden, P., Smith, J. A., Custovic, A., Nilsson, M., & Fowler, S. J., (2013). Breath metabolomic profiling by nuclear magnetic resonance spectroscopy in asthma. *Allergy*, *68*(8), 1050–1056.

Ioannidis, J. P. A., (2005). Why most published research findings are false. *PLoS Med.*, *2*(8), e124.

Ishikawa, S., Sugimoto, M., Kitabatake, K., Sugano, A., Nakamura, M., Kaneko, M., Ota, S., Hiwatari, K., Enomoto, A., Soga, T., Tomita, M., & Iino, M., (2016). Identification of salivary metabolomic biomarkers for oral cancer screening. *Sci. Rep.*, *6*, 31520.

Izquierdo-García, J. L., Peces-Barba, G., Heili, S., Diaz, R., Want, E., & Ruiz-Cabello, J., (2011). Is NMR-based metabolomic analysis of exhaled breath condensate accurate? *Eur. Respir. J.*, *37*(2), 468.

Jianping, G., Yingchang, Z., Yonggang, W., Feng, W., Lang, L., Ping, W., Yong, Z., & Kejing, Y., (2016). Breath analysis for noninvasively differentiating *Acinetobacter baumannii* ventilator-associated pneumonia from its respiratory tract colonization of ventilated patients. *J. Breath Res.*, *10*(2), 027102.

Jones, O. A. H., & Cheung, V. L., (2007). An introduction to metabolomics and its potential application in veterinary science. *Comp. Med.*, *57*(5), 436–442.

Jones, O. A. H., Sdepanian, S., Lofts, S., Svendsen, C., Spurgeon, D. J., Maguire, M. L., & Griffin, J. L., (2014). Metabolomic analysis of soil communities can be used for pollution assessment. *Environ. Toxicol. Chem.*, *33*(1), 61–64.

Jose, L. I. G., Jesús, R. C., Pablo, C., Pilar, F. S., Andrés, E., & José, A. L., (2012). Metabolomic analysis as a diagnostic tool for acute respiratory distress syndrome caused by H1N1 influenza infection in humans. In: *A25. Predicting Development and Outcomes in Acute Lung Injury*, American Thoracic Society, A1149.

Kalluri, U., Naiker, M., & Myers, M., (2014). Cell culture metabolomics in the diagnosis of lung cancer–the influence of cell culture conditions. *J. Breath Res.*, *8*(2), 027109.

Keeler, J., (2005). *Understanding NMR Spectroscopy*. *1 edn.*, John Wiley and Sons: Chichester, ISBN: 0470017872, pp. 459.

Kim, L., Mitrevski, B., Tuck, K. L., & Marriott, P. J., (2013). Quantitative preparative gas chromatography of caffeine with nuclear magnetic resonance spectroscopy. *J. Sep. Sci.*, *36*(11), 1774–1780.

King, J., Mochalski, P., Kupferthaler, A., Unterkofler, K., Koc, H., Filipiak, W., Teschl, S., Hinterhuber, H., & Amann, A., (2010). Dynamic profiles of volatile organic compounds in exhaled breath as determined by a coupled PTR-MS/GC-MS study. *Physiol. Meas.*, *31*(9), 1169–1184.

Koczulla, R., Dragonieri, S., Schot, R., Bals, R., Gauw, S. A., Vogelmeier, C., Rabe, K. F., Sterk, P. J., & Hiemstra, P. S., (2009). Comparison of exhaled breath condensate pH using two commercially available devices in healthy controls, asthma and COPD patients. *Respir. Res.*, *10*, 78.

Kostikas, K., Papatheodorou, G., Psathakis, K., Panagou, P., & Loukides, S., (2003). Oxidative stress in expired breath condensate of patients with COPD. *Chest*, *124*(4), 1373–1380.

Kubáň, P., Kobrin, E. G., & Kaljurand, M., (2012). Capillary electrophoresis – A new tool for ionic analysis of exhaled breath condensate. *J. Chromatogr. A.*, *1267*, 239–245.

Kühnle, M., Kreidler, D., Holtin, K., Czesla, H., Schuler, P., Schaal, W., Schurig, V., & Albert, K., (2008). Online coupling of gas chromatography to nuclear magnetic resonance spectroscopy: method for the analysis of volatile stereoisomers. *Anal. Chem.*, *80*(14), 5481–5486.

Kühnle, M., Kreidler, D., Holtin, K., Czesla, H., Schuler, P., Schurig, V., & Albert, K., (2010). Online coupling of enantioselective capillary gas chromatography with proton nuclear magnetic resonance spectroscopy. *Chirality*, *22*(9), 808–812.

Kumar, S., Huang, J., Abbassi-Ghadi, N., Španěl, P., Smith, D., & Hanna, G. B., (2013). Selected ion flow tube mass spectrometry analysis of exhaled breath for volatile organic compound profiling of esophagogastric cancer. *Anal. Chem.*, *85*(12), 6121–6128.

Kwak, J., Fan, M., Harshman, S., Garrison, C., Dershem, V., Phillips, J., Grigsby, C., & Ott, D., (2014). Evaluation of Bio-VOC sampler for analysis of volatile organic compounds in exhaled breath. *Metabolites*, *4*(4), 879.

Lamote, K., Nackaerts, K., & Van Meerbeeck, J. P., (2014). Strengths, weaknesses, and opportunities of diagnostic breathomics in pleural mesothelioma-a hypothesis. *Cancer Epidemiol. Biomarkers Prev.*, *23*(6), 898–908.

Lanucara, F., Holman, S. W., Gray, C. J., & Eyers, C. E., (2014). The power of ion mobility-mass spectrometry for structural characterization and the study of conformational dynamics. *Nat. Chem.*, *6*(4), 281–294.

Lin, S., Liu, N., Yang, Z., Song, W., Wang, P., Chen, H., Lucio, M., Schmitt-Kopplin, P., Chen, G., & Cai, Z., (2010). GC-MS-based metabolomics reveals fatty acid biosynthesis and cholesterol metabolism in cell lines infected with influenza A virus. *Talanta.*, *83*(1), 262–268.

Lowen, A. C., Mubareka, S., Tumpey, T. M., Garcia-Sastre, A., & Palese, P., (2006). The guinea pig as a transmission model for human influenza viruses. *Proc. Natl. Acad. Sci. USA*, *103*(26), 9988–9992.

Luxon, B. A., (2014). Metabolomics in asthma. *Adv. Exp. Med. Biol.*, *795*, 207–220.

Maiga, M., Choi, S. W., Atudorei, V., Maiga, M. C., Sharp, Z. D., Bishai, W. R., & Timmins, G. S., (2014). *In vitro* and *In vivo* studies of a rapid and selective breath test for tuberculosis based upon mycobacterial CO dehydrogenase. *mBio.*, *5*(2), e00990-14.

Martin, A. N., Farquar, G. R., Jones, A. D., & Frank, M., (2010). Human breath analysis: methods for sample collection and reduction of localized background effects. *Anal. Bioanal. Chem.*, *396*(2), 739–750.

Miekisch, W., & Schubert, J. K., (2006). From highly sophisticated analytical techniques to life-saving diagnostics: Technical developments in breath analysis. *Trends Analyt. Chem.*, *25*(7), 665–673.

Minh, T. D., Oliver, S. R., Ngo, J., Flores, R., Midyett, J., Meinardi, S., Carlson, M. K., Rowland, F. S., Blake, D. R., & Galassetti, P. R., (2011). Noninvasive measurement of plasma glucose from exhaled breath in healthy and type 1 diabetic subjects. *Am. J. Physiol. Endocrinol. Metab.*, *300*(6), 1166–1175.

Montuschi, P., Mores, N., Trové, A., Mondino, C., & Barnes, P. J., (2013). The electronic nose in respiratory medicine. *Respiration*, *85*(1), 72–84.

Montuschi, P., Paris, D., Melck, D., Lucidi, V., Ciabattoni, G., Raia, V., Calabrese, C., Bush, A., Barnes, P. J., & Motta, A., (2012). NMR spectroscopy metabolomic profiling of exhaled breath condensate in patients with stable and unstable cystic fibrosis. *Thorax*, *67*(3), 222–228.

Montuschi, P., Paris, D., Montella, S., Melck, D., Mirra, V., Santini, G., Mores, N., Montemitro, E., Majo, F., Lucidi, V., Bush, A., Motta, A., & Santamaria, F., (2014). Nuclear magnetic resonance–based metabolomics discriminates primary ciliary dyskinesia from cystic fibrosis. *Am. J. Respir. Crit. Care Med.*, *190*(2), 229–233.

Motta, A., Paris, D., D'Amato, M., Melck, D., Calabrese, C., Vitale, C., Stanziola, A. A., Corso, G., Sofia, M., & Maniscalco, M., (2014). NMR metabolomic analysis of exhaled breath condensate of asthmatic patients at two different temperatures. *J. Proteome Res.*, *13*(12), 6107–6120.

Motta, A., Paris, D., Melck, D., De Laurentiis, G., Maniscalco, M., Sofia, M., & Montuschi, P., (2012). Nuclear magnetic resonance-based metabolomics of exhaled breath condensate: methodological aspects. *Eur. Respir. J.*, *39*(2), 498.

Muhlebach, M. S., & Sha, W., (2015). Lessons learned from metabolomics in cystic fibrosis. *Mol. Cell. Pediatr.*, *2*(1), 1–7.

Nicholson, J. K., Lindon, J. C., & Holmes, E., (1999). 'Metabonomics': understanding the metabolic responses of living systems to pathophysiological stimuli via multivariate statistical analysis of biological NMR spectroscopic data. *Xenobiotica*, *29*(11), 1181–1189.

Nobakht, M. G. B. F., Aliannejad, R., Rezaei-Tavirani, M., Taheri, S., & Oskouie, A. A., (2015). The metabolomics of airway diseases, including COPD, asthma and cystic fibrosis. *Biomarkers*, *20*(1), 5–16.

Oliver, S. G., Winson, M. K., Kell, D. B., & Baganz, F., (1998). Systematic functional analysis of the yeast genome. *Trends Biotechnol.*, *16*(9), 373–378.

Paris, D., Maniscalco, M., Melck, D., D'Amato, M., Sorrentino, N., Zedda, A., Sofia, M., & Motta, A., (2015). Inflammatory metabolites in exhaled breath condensate characterize the obese respiratory phenotype. *Metabolomics*, *11*(6), 1934–1939.

Park, H. E., Yang, S. O., Hyun, S. H., Park, S. J., Choi, H. K., & Marriott, P. J., (2012). Simple preparative gas chromatographic method for isolation of menthol and menthone from peppermint oil, with quantitative GC-MS and (1) H NMR assay. *J. Sep. Sci.*, *35*(3), 416–423.

Patel, S., & Ahmed, S., (2015). Emerging field of metabolomics: Big promise for cancer biomarker identification and drug discovery. *J. Pharm. Biomed. Anal.*, *107*, 63–74.

Pauling, L., Robinson, A. B., Teranishi, R., & Cary, P., (1971). Quantitative analysis of urine vapor and breath by gas-liquid partition chromatography. *Proc. Natl. Acad. Sci. USA*, *68*(10), 2374–2376.

Peralbo-Molina, A., Calderón-Santiago, M., Priego-Capote, F., Jurado-Gámez, B., & Castro, M. D., (2016). Metabolomics analysis of exhaled breath condensate for discrimination between lung cancer patients and risk factor individuals. *J. Breath Res.*, *10*(1), 016011.

Pereira, J., Porto-Figueira, P., Cavaco, C., Taunk, K., Rapole, S., Dhakne, R., Nagarajaram, H., & Câmara, J., (2015). Breath analysis as a potential and non-invasive frontier in disease diagnosis: An overview. *Metabolites*, *5*(1), 3.

Phillips, C., Mac Parthaláin, N., Syed, Y., Deganello, D., Claypole, T., & Lewis, K., (2014). Short-term intra-subject variation in exhaled volatile organic compounds (VOCs) in COPD patients and healthy controls and its effect on disease classification. *Metabolites*, *4*(2), 300.

Phillips, M., Cataneo, R. N., Chaturvedi, A., Danaher, P. J., Devadiga, A., Legendre, D. A., Nail, K. L., Schmitt, P., & Wai, J., (2010). Effect of influenza vaccination on oxidative stress products in breath. *J. Breath Res.*, *4*(2), 026001.

Phillips, M., Herrera, J., Krishnan, S., Zain, M., Greenberg, J., & Cataneo, R. N., (1999). Variation in volatile organic compounds in the breath of normal humans. *J. Chromatogr. B.*, *729*(1–2), 75–88.

Pieil, J. D., Stiegel, M. A., & Risby, T. H., (2013). Clinical breath analysis: discriminating between human endogenous compounds and exogenous (environmental) chemical confounders. *J. Breath. Res.*, *7*(1), 017107.

Pijls, K. E., Smolinska, A., Jonkers, D. M. A. E., Dallinga, J. W., Masclee, A. A. M., Koek, G. H., & Van Schooten, F. J., (2016). A profile of volatile organic compounds in exhaled air as a potential non-invasive biomarker for liver cirrhosis. *Sci. Rep.*, *6*, 19903.

Prado, C., Marin, P., & Periago, J. F., (2003). Application of solid-phase microextraction and gas chromatography–mass spectrometry to the determination of volatile organic compounds in end-exhaled breath samples. *J. Chromatogr. A.*, *1011*(1–2), 125–134.

Qian, Z., Penghui, L., Yunfeng, C., Wei, Z., Haidong, W., Jiao, L., Jianhua, D., & Huanwen, C., (2016). Detection of creatinine in exhaled breath of humans with chronic kidney disease by extractive electrospray ionization mass spectrometry. *J. Breath Res.*, *10*(1), 016008.

Rabinowitz, J. D., Purdy, J. G., Vastag, L., Shenk, T., & Koyuncu, E., (2011). Metabolomics in drug target discovery. *Cold Spring Harb. Symp. Quant. Biol.*, *76*, 235–246.

Ritter, J. B., Wahl, A. S., Freund, S., Genzel, Y., & Reichl, U., (2010). Metabolic effects of influenza virus infection in cultured animal cells: Intra- and extracellular metabolite profiling. *BMC Syst. Biol.*, *4*, 61.

Rosias, P. P., Robroeks, C. M., Kester, A., Den Hartog, G. J., Wodzig, W. K., Rijkers, G. T., Zimmermann, L. J., Van Schayck, C. P., Jöbsis, Q., & Dompeling, E., (2008). Biomarker reproducibility in exhaled breath condensate collected with different condensers. *Eur. Respir. J.*, *31*(5), 934.

Sakata, K., Yamada, N., Midorikawa, R., Wirfel, J. C., Potter, D. L., & Martinez, A. G. G., (2001). Inductively coupled plasma mass spectrometer and method. *U. S. Patent 6, 265*, 717.

Santini, G., Mores, N., Penas, A., Capuano, R., Mondino, C., Trove, A., Macagno, F., Zini, G., Cattani, P., Martinelli, E., Motta, A., Macis, G., Ciabattoni, G., & Montuschi, P., (2016). Electronic nose and exhaled breath NMR-based metabolomics applications in airways disease. *Curr. Top. Med. Chem.*, *16*(14), 1610–1630.

Schatz, V., Strüssmann, Y., Mahnke, A., Schley, G., Waldner, M., Ritter, U., Wild, J., Willam, C., Dehne, N., Brüne, B., McNiff, J. M., Colegio, O. R., Bogdan, C., & Jantsch, J., (2016). Myeloid cell–derived HIF-1α promotes control of *Leishmania major*. *J. Immunol., 197*(10), 4034–4041.

Scott-Thomas, A., Epton, M., & Chambers, S., (2013). Validating a breath collection and analysis system for the new tuberculosis breath test. *J. Breath Res., 7*(3), 037108.

Silva, G. D., & Beyette, F. R., (2014). *In Alveolar air Volatile Organic Compound Extractor for Clinical Breath Sampling*, 36th Annual International Conference of the IEEE engineering in medicine and biology society, Chicago, IL, USA, IEEE: Chicago, IL, USA, 5369–5372.

Snowden, S., Dahlén, S. E., & Wheelock, C. E., (2012). Application of metabolomics approaches to the study of respiratory diseases. *Bioanalysis, 4*(18), 2265–2290.

Sofia, M., Maniscalco, M., De Laurentiis, G., Paris, D., Melck, D., & Motta, A., (2011). Exploring airway diseases by NMR-based metabonomics: A review of application to exhaled breath condensate. *J. Biomed. Biotechnol,* 403260.

Soriano, J. B., Abajobir, A. A., Abate, K. H., Abera, S. F., Agrawal, A., Ahmed, M. B. et al., (2017). Global, regional, and national deaths, prevalence, disability-adjusted life years, and years lived with disability for chronic obstructive pulmonary disease and asthma, 1990–2015: A systematic analysis for the Global Burden of Disease Study 2015. *Lancet Respir. Med., 5*(9), 691–706.

Spinhirne, J. P., Koziel, J. A., & Chirase, N. K., (2004). Sampling and analysis of volatile organic compounds in bovine breath by solid-phase microextraction and gas chromatography-mass spectrometry. *J. Chromatogr. A., 1025*(1), 63–69.

Stringer, K. A., McKay, R. T., Karnovsky, A., Quémerais, B., & Lacy, P., (2016). Metabolomics and its application to acute lung diseases. *Front. Immunol., 7*(44).

Sweetlove, L. J., Last, R. L., & Fernie, A. R., (2003). Predictive metabolic engineering: a goal for systems biology. *Plant Physiol., 132*(2), 420–425.

Tewari, V., Nath, G., Gupta, H., Dixit, V. K., & Jain, A. K., (2001). 14C-urea breath test for assessment of gastric Helicobacter pylori colonization and eradication. *Indian J. Gastroenterol., 20*(4), 140–143.

Tweeddale, H., Notley-McRobb, L., & Ferenci, T., (1998). Effect of slow growth on metabolism of Escherichia coli, as revealed by global metabolite pool ("metabolome") analysis. *J. Bacteriol., 180*(19), 5109–5116.

Ueta, I., Mizuguchi, A., Okamoto, M., Sakamaki, H., Hosoe, M., Ishiguro, M., & Saito, Y., (2014). Determination of breath isoprene and acetone concentration with a needle-type extraction device in gas chromatography–mass spectrometry. *Clin. Chim. Acta, 430,* 156–159.

Urban, S., & Dias, D. A., (2013). NMR spectroscopy: structure elucidation of cycloelatanene a: A natural product case study. In: *Metabolomics Tools for Natural Product Discovery: Methods and Protocols*, Roessner, U., Dias, A. D., eds. Humana Press: Totowa, NJ, 99–116.

Van Mastrigt, E., De Jongste, J. C., & Pijnenburg, M. W., (2015). The analysis of volatile organic compounds in exhaled breath and biomarkers in exhaled breath condensate in children – clinical tools or scientific toys? *Clin. Exp. Allergy., 45*(7), 1170–1188.

Vries, R. D., Brinkman, P., Schee, M. P. V. D., Fens, N., Dijkers, E., Bootsma, S. K., Jongh, F. H. C. D., & Sterk, P. J., (2015). Integration of electronic nose technology with spirometry: validation of a new approach for exhaled breath analysis. *J. Breath Res.*, *9*(4), 046001.

Wang, C., & Sahay, P., (2009). Breath analysis using laser spectroscopic techniques: Breath biomarkers, spectral fingerprints, and detection limits. *Sensors, 9*(10), 8230.

Wang, H., Liu, X., Nan, K., Chen, B., He, M., & Hu, B., (2017). Sample pre-treatment techniques for use with ICP-MS hyphenated techniques for elemental speciation in biological samples. *J. Anal. At. Spectrum, 32*(1), 58–77.

Wilson, A., (2014). Electronic-nose applications in forensic science and for analysis of volatile biomarkers in the human breath. *J. Forensic Sci. Criminol., 1*(S103), 1–21.

Zhan, X., Duan, J., & Duan, Y., (2013). Recent developments of proton-transfer reaction mass spectrometry (PTR-MS) and its applications in medical research. *Mass Spectrom. Rev., 32*(2), 143–165.

Zhao, Y., Peng, J., Lu, C., Hsin, M., Mura, M., Wu, L., Chu, L., Zamel, R., Machuca, T., Waddell, T., Liu, M., Keshavjee, S., Granton, J., & De Perrot, M., (2014). Metabolomic heterogeneity of pulmonary arterial hypertension. *PLoS One, 9*(2), e88727.

Zhu, J., Bean, H. D., Wargo, M. J., Leclair, L. W., & Hill, J. E., (2013a). Detecting bacterial lung infections: in vivo evaluation of in vitro volatile fingerprints. *J. Breath Res., 7*(1), 016003.

Zhu, J., Jimenez-Diaz, J., Bean, H. D., Daphtary, N. A., Aliyeva, M. I., Lundblad, L. K., & Hill, J. E., (2013b). Robust detection of *P. aeruginosa* and *S. aureus* acute lung infections by secondary electrospray ionization-mass spectrometry (SESI-MS) breathprinting: from initial infection to clearance. *J. Breath Res., 7*(3), 037106.

PART III
Computational Tools

CHAPTER 8

THE NEED OF EXTERNAL VALIDATION FOR METABOLOMICS PREDICTIVE MODELS

RAQUEL RODRÍGUEZ-PÉREZ,[1,2] MARTA PADILLA,[1] and SANTIAGO MARCO[1,2]

[1]*Signal and Information Processing for Sensing Systems, Institute for Bioengineering of Catalonia, Baldiri Reixac 4–8, 08028-Barcelona, Spain*

[2]*Department of Engineering: Electronics, Universitat de Barcelona, Martн i Franquйs 1, 08028-Barcelona, Spain,* *E-mail: smarco@ibecbarcelona.eu*

ABSTRACT

Over the last decade, metabolomics research has produced thousands of research works. Many of them were describing the ability of machine learning methods to detect diverse health conditions based on mass-spectrometry, nuclear magnetic resonance or artificial olfaction analysis of body fluids.

While few success stories exist, most described applications never found the road to clinical exploitation. Most described methodologies were not reliable and were plagued by numerous problems that prevented practical application beyond the lab. This work gives some insight on the reasons behind this lack of generalizability and emphasizes the need of external validation in metabolomics research. We describe some statistical and methodological pitfalls of the current data analysis practice and we give some best practice recommendations for researchers in the field.

8.1   INTRODUCTION

Metabolome refers to the complete set of low weight compounds found within a biological sample or organism (Oliver et al., 1998). As metabolites are substrates and products of cellular activity, they reflect an overview of the different biological processes that occur in the body. Therefore, metabolomics studies the final downstream product (Fernie et al., 2004; Monteiro et al., 2013), which is determined by the interaction between genome and environment (including age and lifestyle) (Gebregiworgis and Powers, 2012; Worley and Powers, 2013). It is said that the metabolome provides a dynamic and sensitive measure of the phenotype (Lämmerhofer and Weckwerth, 2013). Moreover, it is the smallest domain, with approximately 5,000 metabolites, versus the 3.2 billion DNA bases and 35,000 coding genes of the genome and more than 100,000 proteins of the proteome. However, it is a more diverse domain and more physically and chemically complex than the other 'omes' (Dettmer et al., 2007).

Metabolomics involves the comparison of metabolomes between groups of samples to find differences in their profiles (Agilent, 2012). Those differences may be correlated to a disease being studied or changes in metabolic output in toxicology studies when a drug candidate is introduced to a test subject, as studied in pharmaceutical industry.

Metabolomics data can be obtained from different cells, tissues, organs or body fluids. Sampling of many biological samples of the latter category is minimally or non-invasive for the patient. Biofluids samples integrate biological processes that occur in cells, tissues and organs with which the biofluid interacts (Lämmerhofer and Weckwerth, 2013). Some examples are blood serum and plasma, urine, faeces, sweat, saliva or breath. From the headspace of these matrices, volatile organic compounds (VOCs) can be collected. VOCs are organic analytes with a substantial vapour pressure (typically analytes less than 300 Da can be considered volatile compounds) and their analysis is a promising way of monitoring human chemistry and health related conditions. Since VOCs are produced by specific biochemical pathways in the body, altered physiological processes as a consequence of disease can cause production of new VOCs or alteration of the normal levels (Schmidt and Podmore, 2015). Therefore, VOCs analysis has the potential to enhance personalized medicine (Broza and Haick, 2014). In fact, many patterns of VOCs have already been reported to be correlated with diseases or pathologies such as chronic obstructive pulmonary disease (COPD) (Berkel et al., 2010), lung cancer (Bajtarevic et al., 2009), gastric lesions and gastric cancer (Amal et al., 2015), colorectal cancer (Altomare et al., 2013), urological malignancies (Aggio et al., 2016) or melanoma (Abaffy et al., 2013).

Some of the most popular and powerful instrumental techniques are nuclear magnetic resonance (NMR) spectroscopy, and mass spectrometry (MS) coupled with a previous separation method such as gas chromatography (GC) or liquid chromatography (LC) (Strathmann and Hoofnagle, 2011). These are well-established techniques for the generation of metabolic profiles and each of them has strengths and weaknesses. For instance, GC or LC coupled with MS (also known as GC-MS and LC-MS, respectively) need more sample preparation than NMR. Nevertheless, they also have higher sensitivities, which allow identification of metabolites in lower abundance. In addition, GC-MS is used when the metabolites of interest have low molecular weight or are volatile (Lämmerhofer and Weckwerth, 2013), for example, in breath analysis.

Electronic noses (ENs) or artificial olfaction systems (AOSs) (Halsey et al., 2015) is another important instrument that has been employed for the analysis of several body fluids in health-related applications (D'Amico et al., 2008; Gardner et al., 2000). AOSs are composed by arrays of chemical sensors, which show intrinsic partial selectivity. Thus, AOSs also rely on signal processing in order to increase selectivity (Marco and Gutierrez-Galvez, 2012).

In most occasions, metabolomic analysis is framed as a biomarker discovery study. Biomarkers are expected to be used for screening, diagnosis, personalized therapies, monitoring and prognosis. They are also important in drug discovery and development to understand the mechanisms of action of a drug, or to investigate its toxicity.

For biomarker discovery, we intend to acquire the global metabolic profile of a sample, which is known as untargeted metabolomics or metabolic fingerprinting (Gosho et al., 2012). Technology advances in analytical instrumentation have provided the possibility of examine hundreds or thousands of metabolites in parallel. Therefore, current instrumentation, for example, MS, leads to complex data sets. Rather than variations in a single parameter, the pathology changes the physiological status and the relative abundance of metabolites (Tautenhahn et al., 2012). Thus, this discipline aims to simultaneously monitor as many metabolites as possible from biological samples without bias (Patti et al., 2012), to further compare metabolic fingerprints among distinct experimental groups. Hence, analytical techniques require low detection limits and good selectivity (Bussche et al., 2015), but also a great coverage of the metabolome in a single analysis.

In untargeted metabolomics, current analytical instrumentation offers the possibility to have a semi-quantitative evaluation of hundreds to thousands of metabolites per sample. However, the cost per analysis and mostly the difficulties to recruit proper cohorts for the condition and control groups produce in general a limited number of samples per class. In consequence, these techniques give high-dimensional data and studies are often characterized by small sample size (Marco, 2014; Vinaixa et al., 2012). From the obtained data sets covering hundreds of metabolites, computational techniques for biomarker discovery try to identify which ones are discriminants between the control and case groups. In fact, health systems worldwide increasingly require more point-of-care attention and testing a small count combination of biomarkers would be preferred over measuring hundreds of them (Xia et al., 2013). Due to the high dimensionality of the data sets and the limited number of samples, this computational task is not exempt from important difficulties (Azuaje, 2010; Kulasingam and Diamandis, 2008; Larrañaga et al., 2006).

In brief, signal processing and data analysis for analytical instrumentation, also known as chemometrics or bioinformatics in this context, are of increasing importance in omics as well as in analytical chemistry and chemical sensing or detection. Current chemical instrumentation offers enormous capabilities for signal recording and storage, which might cause that data analysis and interpretation become the bottleneck in the process (Lämmerhofer and Weckwerth, 2013). In many studies, such as biomarker discovery studies, the goal of the data analysis process is to detect changes in the metabolome between two biological condition groups, usually a case group and a control group of sample profiles (Azuaje, 2010).

From a computational perspective, the process of building predictive models from omics data can be considered as a standard pattern recognition problem, the general workflow of which is shown in Figure 8.1.

Even though biomarker discovery may be based on univariate hypothesis testing, multivariate predictive models together with feature selection methods are frequently employed in metabolomics studies (Smolinksa et al., 2012). Predictive models built with the selected biomarker candidates can provide diagnostic or prognosis information. Additionally, models may be the basis for ordering additional tests or selecting therapies. Thus, the correct assessment of model validity is of utmost importance.

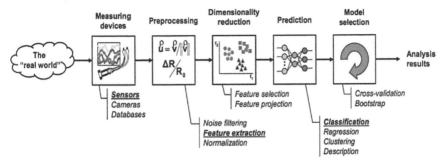

FIGURE 8.1   (See color insert.) General workflow of a bioinformatic solution or pattern recognition problem. First, measuring devices are needed (e.g., NMR, GC-MS, LC-MS) and the obtained data is pre-processed. The number of features extracted from the data (data dimensionality) is often reduced due to its large value with respect to the often small sample size. Then, a qualitative or quantitative prediction model is built with a set of known samples (training set) and evaluated (validation set) to both select the optimum model and estimate the performance (From Nagle et al., (1998). *IEEE Spectrum., 35*(9), 22–31. With permission).

Feature selection methods are usually applied to identify a subset of variables that has predictive power for the condition under study (Saerys et al., 2007). In any case dimensionality reduction (Figure 8.1) through feature selection is a key block in data processing for metabolomics to avoid the problems of the curse of dimensionality (Bishop, 2006; Saerys et al., 2007).

Numerous studies and research findings have been published in the domain of biomarker discovery in omics sciences (Bussche et al., 2015; Dettmer et al., 2007; Machado and Laskowski, 2005; Schmekel et al., 2014; Westhoff et al., 2011). However, the generalized lack of reproducibility of the results indicates that methodological problems plague these studies. As a result, the vast majority of proposed biomarkers in literature have never found clinical application (Ducker and Krapfenbauer, 2013).

In many occasions, when results cannot be reproduced by other groups, statistical bias is proposed to explain the original overoptimistic results. In fact, some reported biomarkers providing perfect predictive accuracy have been later reported to be false and results attributed to bias. A popular example was a study that claimed to have found proteomic patterns in serum to identify ovarian cancer, but it was posteriorly demonstrated that observed differences were due to distinct sample storage times (Petricoin et al., 2002). This reflects the presence of some limitations in the field and some authors have already pointed out the problem of irreproducible science, and in particular bad statistical practices (Hayden, 2013; Ioannidis, 2005; Xia et al., 2013).

In this work, we review some issues that usually degrade model validity. Although there are several difficulties in building proper predictive models, here we focus on analytical robustness and the problem of weak statistics or computational methodology. Analytical robustness is discussed in Section 8.2, while in Section 8.3 some errors committed in the evaluation of predictive models are described. Specifically, Section 8.3 discusses the main limitations in the analysis of metabolomic data, which are related to hypothesis testing and p-value misunderstanding, confounding factors, and weak validation strategies.

8.2   ANALYTICAL ROBUSTNESS

Reproducibility problems are in part due to the existence of a validity domain of a data model. In many occasions, authors assume that models will work outside this domain of validity, or they simply ignore that the model validity is not guaranteed beyond the instrumental and sampling conditions of the study.

The first condition for an extended validity of a data model is that the analytical procedure has to be robust. Robustness (or ruggedness) of an analytical procedure has been defined by the World Health Organization as "the ability of the procedure to provide analytical results of acceptable accuracy and precision under a variety of conditions. The results from separate samples are influenced by changes in operational or environmental conditions" (WHO, 2007).

Robustness should be evaluated along the entire analytical procedure, including sampling, instrumental analysis and data processing. The key for robustness evaluation is to introduce forced variations on the analytical procedure and check if the results are still consistent. The most common reasons for lack of robustness are: instrumental shifts in time, hypersensitivity to instrument conditions, operation protocol, sampling, and presence of manual (not automated) steps.

Among them, we would like to remark that all instruments shift over time (Figure 8.2) and this is an important issue to take into account (Marco, 2014). For instance, chromatography might present instrumental variations due to column aging, temperature effects or changes on ionization efficiency, which can make intensity value measurements change

(Fernández-Albert et al., 2014; Lämmerhofer and Weckwerth, 2013; Zelena et al., 2009). In the same way, chemical sensors are affected by ageing due to the degradation of the chemo-physical properties of their sensing material. This effect also results in a slow change of the sensors' response along time (Artursson et al., 2000; Padilla et al., 2010; Romain et al., 2002). Figure 8.3 shows a Principal Component Analysis (PCA) scoreplot of urine analysis by LC-MS, which presents a clear drift due to day of analysis. If these instrumental shifts are not properly considered during experimental design and posterior analysis, time can become a confounding factor (Fernández-Albert et al., 2014).

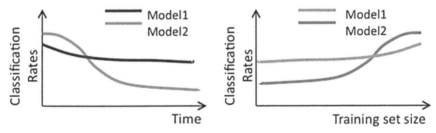

FIGURE 8.2 Illustration of the robustness of two data models with respect to time and training set size: (a) the performance of model 2 is more stable in time than model 1, then model 2 is more robust with respect to time. (b) model 1 is more robust than model 2 because its performance is stable with respect to the training set size, although slightly worse than model2 for larger training sets (From Padilla et al., (2013). *Improving the Robustness of Odor Sensing Systems by Multivariate Signal Processing*. In: *Human Olfactory Displays and Interfaces: Odor Sensing and Presentation*; Nakamoto, T., (Ed.). *IGI Global: Hershey,* pp. 296–316. With permission.).

The sensitivity of the analytical procedure to each metabolite has to remain stable because most biomarker discovery techniques rely on semi-quantitative evaluation of metabolites composition. Indeed, keeping the sensitivity of the analytical procedure stable is a hard challenge, specially for untargeted studies for which key metabolites to be detected are not known *a priori*. Proposed solutions include the application of calibration routines based on external or internal standards to detect and counteract analytical variation, when it is possible. In general, it is recommended to obtain a quantitative measure of the variation within and between experiments, which can be done by means of quality control (QC) samples (Kamleh et al., 2012). QC samples are special substances measured and treated as usual samples under study, thus exposed to the same operating conditions. They are regularly presented to the instrument always in known and constant amounts. The substances that compose these QC samples can be commercial, stable materials or pooled, i.e., a mix of equal aliquots from a representative set of the study samples. In fact, the formulation of these samples is not always obvious. For example, for the analysis of urine (Chan et al., 2011), the use of a polled mixture of all samples in the study was proposed. The rational was to have a sample that contained all the metabolites present in the samples of the study.

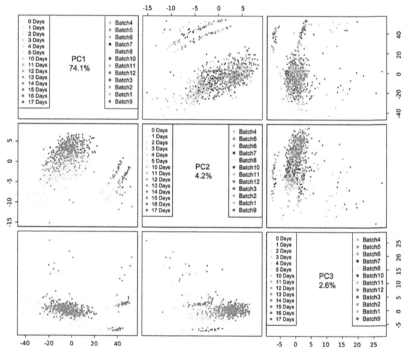

FIGURE 8.3   (See color insert.) PCA scoreplot of urine samples analysed by LC-MS organized either by batch or day of measurement. The numbers in the diagonal plot represent the variance captured by each principal component (From Fernández-Albert et al., (2014). *Bioinformatics., 30,* 2899–2905. With permission.).

The inclusion of the QC samples in the analytical process allows the evaluation of the reproducibility and the repeatability of the analytical procedure, and it is strongly recommended when drifts are expected. QC samples are the base of the QC protocols which also include: selection of the time between QC process applications, QC samples material, counteraction methods (e.g., by signal processing) and decision taking regarding data rejection and instrument maintenance (Dunn et al., 2012). See Figure 8.4 for an illustration of the role of calibrant samples. QC processes ensure that the data model is valid in the domain of the current measurement system. However, the model might not have generalization capability for another set of conditions in which QC samples are not available.

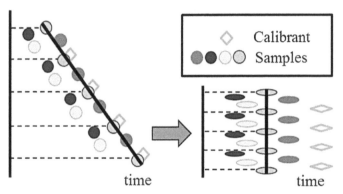

FIGURE 8.4   Illustration of a linear correction of time shift by including specific QC samples (calibrant) equally distributed in time along the whole experiment.

An interesting approach to guarantee robustness and model validity beyond a particular laboratory is multicentre studies (Bijlsma et al., 2006). Nevertheless, this approach is usually encountered in posterior phases regarding the validation of biomarkers. Most published findings in this area remain limited to a single laboratory/hospital with small cohorts.

8.3   ERRORS IN THE EVALUATION OF PREDICTIVE MODELS

In this section we will review some statistical issues that unless handled correctly may degrade the quality of the data analysis and consequently the interpretation of the results.

There are some common pitfalls when evaluating predictive models. Since rigorous methods of data analysis are vital to produce valid conclusions, we discuss how to avoid these errors and how to design robust methodologies for data analysis. More specifically, we review three of the most frequent errors of data processing in this field: (i) misinterpretation of p-values, (ii) confounding factors, and (iii) overfitting.

8.3.1 MISINTERPRETATION OF P-VALUES

Many studies focus on finding differences in metabolome composition between two (biological) condition groups through hypothesis testing, which relies on the computation of a single index per metabolite, the p-value. Ronald Fisher introduced the p-value in the 1920s as an informal way to judge whether results were consistent with what random chance may produce (Goodman, 1999b; Nuzzo, 2014). Now, p-value is frequently used as decision rule taking into account the arbitrary convention of 95% of confidence.

In hypothesis testing for metabolomics, p-value is used as a univariate measure of significance of differential abundance levels of metabolites between two groups of samples. For a given metabolite, the null hypothesis of equal abundance level between a case and a control group is rejected when the p-value is lower than a pre-selected value (significance level), usually 5% or 1%.

Applying the same statistical test to each of the metabolites in a data set would result in a high number of false positives (also known as false discoveries or type I errors). Therefore, to reduce the probability of such error type, methods for multiple testing have been developed. These methods apply hypothesis testing considering the whole set (or a subset) of the measured metabolites, i.e., the whole set of tests or hypotheses. For this, several strategies have been followed, such as controlling the family-wise error rate (FWER) or the false discovery rate (FDR) (Benjamini and Hochberg, 1995). FWER is the probability of having at least one false positive among all the tests, whereas FDR is the expected proportion of false positives among all the significant tests. The latter has been especially conceived for data sets with large number of tests and small sample size, as it is the case in genomics. Bonferroni correction (Bland and Altman, 1995) is a method that follows the FWER strategy, while the p-value step-up method proposed by Benjamini and Hochberg (Benjamini and Hochberg, 1995) controls FDR. Moreover, other quantities derived from the mentioned strategies, such as the q-value (Storey, 2003) or the local FDR (LFDR) (Efron et al., 2001), can be used analogously to p-value for multiple comparisons, even for data sets with small number of tests (Bickel, 2013; Padilla and Bickel, 2012).

Aside multiple hypothesis testing, the effect size (ES) of the parameter of interest is an important variable for data analysis. In metabolomics, the ES can be given by the strength of the differential abundance levels of the metabolites between the two condition groups, i.e., the difference between the means of both groups. The p-value by itself does not provide information about the ES of the metabolite under study. It may occur that a metabolite showing small ES is found statistically significant. Indeed, p-value (and also ES) depends on the sample size and it can be extremely unstable with a small number of samples (Halsey et al., 2015), as shown in Figure 8.5. Furthermore, an important effect may also be hidden due to a low number of studied subjects (Gardner and Altman, 1986). Some studies consider the ES along with the p-value to decide whether a metabolite is significant or not. For this, they establish one (or two) threshold for the estimated ES beyond which the metabolite is a candidate for significance. Then, such metabolite is found significant if it also has a small p-value. This means that two or three thresholds have to be pre-selected; one for p-value (usual 5%), and one or two for the ES. A 'volcano plot' representing the p-value (or equivalent quantity derived from multiple hypothesis testing) versus the ES, helps to visualize the regions of significant metabolites according to the three pre-selected thresholds (Patti et al., 2012, 2013) (Figure 8.6).

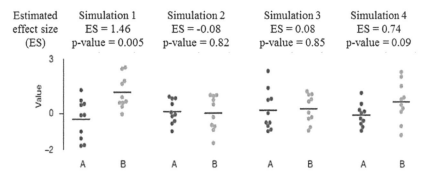

FIGURE 8.5 Illustration of the instability of the p-value for small sample sizes (Halsey et al., 2015). For each out of four simulations, 10 values are randomly selected from each of two populations normally distributed A~N(0,1) and B~N(0.5,1). The estimated ES, which is the difference between the means of each sample population, is represented by a horizontal line and its value is shown along with the p-value. Notice that the estimated ES varies around the true ES (=0.5) and that the range of variation of the p-values is quite large. Simulation 1 would determine that the feature is statistically significant, but the estimated ES has large error, while remaining simulations result in non-significant p-values and estimated ESs closer to the true one. (From Halsey, et al., (2015). *Nat. Methods, 12,* 179–185. With permission.).

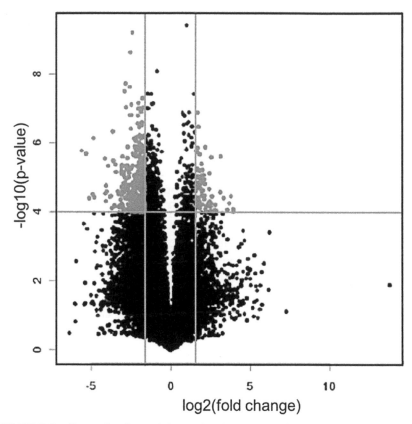

FIGURE 8.6 **(See color insert.)** Example of a volcano plot. The lighter samples are selected, since the value of their p-value is lower than a stated threshold and their ES (in the form of a fold change) is located outside the region defined by another threshold, specifically ±ES threshold (From Patti et al., (2012). *Nat. Rev. Mol. Cell. Biol., 13,* 263–269. With permission.).

In addition, a metabolite found to have high ES between the two condition groups might not provide a correct classification of phenotypes, even though it is statistically significant. Confusing statistical significance with predictive accuracy is a typical error. Height difference between genders is a simple example that reflects the difference between both. In general, males are taller than females, and this statement could easily reach statistical significance in a hypothesis test even for moderate sample sizes. Still, a model to predict gender only based on height will provide a large rate of misclassifications. Therefore, characterizing the performance of a given metabolite by means of a p-value under a certain hypothesis testing does not inform about its suitability as a biomarker (Broadhurst and Kell, 2006; Marco, 2014). Nevertheless, it is quite common to focus research on reaching statistical significance and confound p-value with medical relevance or predictive ability.

The obsessive use of the p-value in metabolomics statistics is plagued with additional problems. Recently, some scientists have popularised the term "p-hacking" (Nuzzo, 2014). It refers to try multiple things, even unconsciously, until the desired result (i.e., a significant p-value) is reached. According to them, "p-hacking" is common in disciplines that search for small effects which may be hidden in noisy data.

From the previous discussions on the problems of the p-value for hypothesis testing, we recommend to take an alternative path. Instead of focusing on hypothesis testing, we strongly recommend to build predictive classification models to assess whether putative biomarkers have predictive performance or, in other words, if the sensitivity and specificity of the prediction satisfies clinical demands of use. Additionally, the use of the Area Under the Curve ROC curve (AUC) as figure of merit provides a fuller characterization of the performance of the predictive model, avoiding the influence of the classification threshold.

The validation of these predictive models typically relies on cross-validation (CV) methods (Section 8.3). Moreover, a proper validation of the statistical significance of the figures of merit obtained requires implementing a permutation test. A permutation test with a sufficient number of repetitions can be used to compute a p-value for the obtained figure of merit. Thus, it allows to check whether finite sample effects could lead to those number by pure chance. The p-value from a permutation test provides additional knowledge about the evidence of the results but it should not lead to a binary thinking (i.e., significant versus no significant) nor to be taken as an absolute truth (Leek and Peng, 2015; Marco, 2014).

On the other hand, Bayesian Statisticians have been extremely critic against the p-value use. Among other criticisms, they claim that the current use of the p-value completely ignores the prior odds of the null hypothesis. In plain words, we may say that the more a hypothesis challenges the current scientific understanding, the larger has to be the evidence. Instead, they propose to use a combination of the prior odds with the Bayes Factor (Goodman, 1999b) (*see,* Goodman (1999a) for further discussions). Alternatively, if prior probabilities are not known, they can be estimated from the data itself. Some predictors based on empirical Bayes approaches for large-scale problems are discussed by Efron (2009). They mostly rely on multiple hypothesis testing and estimation of the metabolites' ESs. In any case, the more implausible the hypothesis, the greater the chance of a false alarm, independently of the p-value (Nuzzo, 2014).

8.3.2 CONFOUNDING FACTORS

Confounding factors are features that are not equally distributed between the classes under study and can be the underlying reason for the observed differences (Broadhurst and Kell, 2006; Kukull and Ganguli, 2012). Confounders might be correlated with another uncontrolled variable, which truly underlies the variance of the feature (Greenland and Morgenstern, 2001; Miekisch et al., 2012). Figure 8.7 illustrates a simple example in which gender becomes a confounder, causing different metabolic fingerprints, which could be mistakenly associated to the disease.

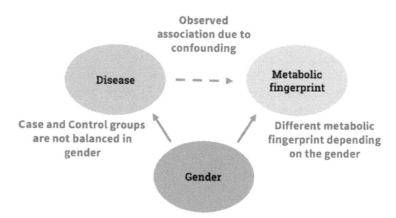

FIGURE 8.7 Confounding factor example. Gender affects to the metabolic fingerprint of case and control groups, and those groups are not balanced according to the gender. Therefore, an association between the condition or disease under study is observed due to the confounding factor.

Indeed, confounding factors are one of the major limitations and causes of bias. Ransohoff (Ransohoff, 2005) stated that "bias is the most important threat to validity" and "all observational research should be presumed guilty of bias, until proven innocent." Such confounding variables causing bias might be due to factors associated with patients, but they may also be introduced at any stage along the experimental procedure (Lämmerhofer and Weckwerth, 2013). Thus, confounding variables can be divided into two types:

i) *Clinical*, such as gender, age, body mass index, smoking status, comorbidities or treatments.
ii) *Instrumental*, such as different location, operator, sampling or storage conditions, temperature, humidity or instrumental (Marco, 2014).

Potential confounders might be prevented by means of a careful experimental design. To be sure that the changes come from the expected cause, the rest of the factors should play no role. Let us consider one example from literature. McLerran et al. (2008) aimed to validate the ability to detect prostatic cancer of serum proteomic profiling using surface-enhanced laser desorption/ionization time-of-flight mass spectrometry (SELDI-TOF-MS). However, they found that their proposed method was unable to differentiate between prostate cancer, benign prostatic hyperplasia, and control groups of samples due to bias. Such bias mainly came from the data collection protocol. All samples from prostatic cancer were collected in 1995 but one in 1996, and all control samples were collected in 1996. Data analysis revealed a clear distinction between the two groups under study. Nonetheless, were they discriminating cancer or the time of sample collection?

Although it is practically impossible to control every factor, there are some strategies to reduce confounders at design stage (Jepsen et al., 2004; Pourhoseingholi et al., 2012):

- *Randomizing*. Experiments are randomized for expected confounders that can be controlled. This is mainly to reduce experimental or instrumental confounders, such as time, operator, location, machine.
- *Restriction*. Some conditions are specified and remained constant during all the experiments. For instance, the same operator, location, and instrument. This might cause that results are local to these conditions.
- *Matching*. Groups are matched so that they are balanced for potential confounders. To do so, it is necessary to dispose of patients' metadata. Nevertheless, sometimes it is complicated to find matched groups.

If confounder's control is not correctly performed or some factors cannot be controlled with experimental design, confounding factors must be taken into account during the data analysis. There are three main approaches to reduce confounders during data analysis (Pourhoseingholi et al., 2012):

i)   *Groups stratification.* The sample groups involved in the study (case and control) are divided in groups in which potential confounders are constant and the analysis is performed separately. This approach may be impossible to perform when the number of samples is limited.

ii) *Multivariate analysis.* Variance explained by confounding factors may be characterised by introducing them in predictive models. Moreover, this variance could be subtracted from the original data with component removal methods, such as Component Correction (Artursson et al., 2000) or Common PCA (Flury, 1984).

iii) *External validation.* Bias could be different in an external set of samples. Therefore, including blind samples for model validation will combat bias as well.

From our point of view, external validation is the ultimate technique to understand and explore the validity of the predictive model. For instance, external samples could have been collected in a different time batch, with a different instrument or in a different hospital. Unfortunately, external validation is not prevalent in most published works in the area of metabolomics. In the fraction of papers where external validation is implemented, samples are often extracted from the same data set, consequently they have been obtained under the same conditions. In these cases, external validation does not serve the purpose of exploring model validity beyond the conditions of the study.

8.3.3   OVERFITTING CONTROL THROUGH VALIDATION TECHNIQUES

Overfitting consists in modelling noise instead of information. It is produced when a learning algorithm is overtrained to a particular set of data and starts to model their intrinsic noise. Hence, the resultant model will be complex and particularly fitted to correctly predict the samples used to build it. Consequently, said model is optimized to ease prediction of a particular set of samples and it cannot generalize properly. Figure 8.8 shows an example of two models aiming to classify artificial data into two classes (A and B). It illustrates a comparison between a simple model, which does not provide a perfect classification of all the training samples, and a model that fits perfectly to all data points. The simpler model would probably have more generalization capacity.

Overfitting occurs when the sample count is smaller or of the same order of magnitude as the model degrees of freedom. Linear discriminant analysis (LDA) is a classic example of a technique prone to overfitting. Still today LDA scoreplots in training are shown in archival journals as proof of discrimination even when dimensionality is much higher than the number of samples. However, it is well known that even data from the same distribution might seem separated with small sample to dimensionality ratios in LDA scoreplots (Qiao et al., 2008).

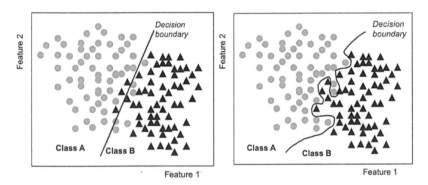

FIGURE 8.8 Illustration of overfitting. Scoreplot of two models aiming to classify an artificial data set into two classes A and B; a simple model (left) and a complex model (right) that perfectly fits the data (overfitting) (adapted from Duda et al., (2000). *Pattern Classification,* 2nd ed. John Wiley and Sons: New York).

Aside from LDA, Partial Least Squares – Discriminant Analysis (PLS-DA) models are also likely to overfit (Szymanska et al., 2011; Westerhuis et al., 2008). Again, PLS-DA scores plots in training are often seen in medical or bioinformatics journals (Brereton and Lloyd, 2014). PLS-DA scoreplots can achieve a perfect separability even with random data with arbitrarily assigned classes are coming from the same distribution, as Figure 8.9 shows.

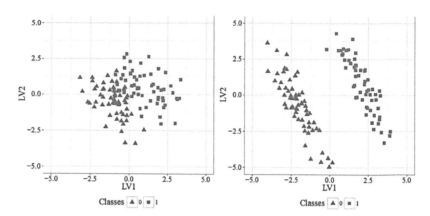

FIGURE 8.9 PLS-DA scoreplots in training for simulated data coming from a Normal distribution N(0,1). Cases of 120 samples and (a) 128 features, (b) 512 features. Notice that groups are more artificially differentiated as number of features increases, even though the number of samples and the distribution of both groups are the same in both cases.

Overfitting is one of the principal factors by which some studies present biomarkers with high performance but not reproducible results (Baldi and Brunak, 2001). This problem can be controlled by strict validation procedures (Brereton, 2006; Broadhurst and Kell, 2006; Marco, 2014). Validation is an essential part of a methodology of data analysis and it has two main purposes: (i) complexity optimization, and (ii) performance estimation of a model.

i)   *Complexity optimization.* Most classifiers have a complexity parameter (model order) that allows to control their flexibility, that is their ability to fit the underlying order. The "k" parameter in k-Nearest Neighbors (k-NN), the number of latent variables (LV) in PLS-DA, or the number of hidden neurons in a multi-layer perceptron are examples of parameters that need to be optimized.

ii)   *Performance estimation.* The generalization capability of a model must be assessed by evaluating its predictive performance.

Validation is based on considering different data partitions for training and testing the model. CV is a widely applied strategy that can be implemented in a number of ways: hold-out, k-fold, leave one out, random subsampling (also named Monte Carlo CV) (Xu et al., 2004) or bootstrap (Efron, 1983; Efron and Gong, 1983; Efron and Tibshirani, 1997).

The method that uses a training subset to build a model, and a validation set both to optimize its complexity and estimate its performance is called 'internal validation.' Actually, internal validation consists in reporting the performance of the optimum model for a given subset of test data. Therefore, even though the estimated predictive power was high, the model might fail to predict new samples. It has been shown that performance assessment by internal CV is not enough (Braga-Neto and Dougherty, 2004; Molinaro et al., 2005). When model complexity optimization and performance estimation tasks are made using the same subset, the evaluation of the model performance is overoptimistic. To ascertain whether the obtained accuracy has been obtained by chance, permutation tests are a must (Ojala and Garriga, 2010).

The performance with new blind samples is worse than expected based on the estimated performance in internal validation. For this reason, it is more rigorous to estimate performance on blind samples, also named 'external validation' (Steyerberg et al., 2003) (Section 8.2). This is probably the best way to check the generalizability of the model.

According to the discussion above, the recommended model building strategy would be a three-way data split, which works as follows. Total data set is split in three subsets (Figure 8.10). The terminology may vary according the science domain, but we will call them training, internal validation and external validation subsets. Using the training and internal validation subsets, model complexity is scanned in a predetermined range of the complexity controlling parameter. Models' parameters are set in the training set, whereas the model's performance is estimated using the internal validation subset. Model complexity is selected as the one that minimizes errors in internal validation. Once complexity is decided, training and internal validation subsets are fused (known as calibration subset) and the final model with optimum parameters is built. The performance of the final model is estimated with the external validation subset, which is then constituted by samples that are blind for the model.

Although the necessity of external validation has been highly recommended by some authors (Boulesteix, 2009; Varma and Simon, 2006; Westerhuis et al., 2008), many studies only report internal validation results (Schmekel et al., 2014). Irreproducibility is strongly related to the current practices in model validation, since external validation is rare in metabolomics field (Ghosh and Poisson, 2009). For instance, about a quarter of published papers in cancer detection through breath analysis did not perform any model validation, whereas the rest only did internal validation (Krilaviciute, 2015).

This is chiefly due to small sample size and no perception of overfitting risk when data examples are used both to optimize and test a data model. For example, a common error is to select the features (biomarkers) using all the samples and later splitting the data set containing only these features into training and test subsets (Aggio et al., 2016; Berkel et al., 2010). While from that point, the methodology could be apparently correct, the initial feature selection before any data partition leads to a strong overfitting. Despite the posterior partition, this methodology is using the same samples for model building (feature selection) and model evaluation.

Usually small sample conditions are argued to avoid external validation. 'Double-cross validation' (Figure 8.10) appears to be a useful validation strategy for small sample size studies (Filzmoser et al., 2009; Westerhuis et al., 2008). In double CV, both internal and external validations are performed with CV methods. The main advantage is that all samples are used as blind for performance estimation. A drawback of this type of validation is that time effects are not considered and calibration samples may be simultaneously in the past and in the future of the blind samples. This is not representative of the final use of the model (posterior to the model building phase) and may provide overoptimistic results in the presence of time drifts. Anyhow, double CV provides an accurate estimation of model performance in new samples even when the number of samples is limited.

8.4   CONCLUSIONS

In this chapter we have reviewed some of the methodological problems that plague metabolomics research and underlie in part the lack of reproducibility of published works.

We have reviewed the importance of analytical robustness. If results are too sensitive to the instrumental conditions, it will be really difficult to replicate the results, either by the same team later in time or by independent teams. Untargeted analysis is a challenge due to the parallel measurement of hundreds of metabolites. Since metabolomics data processing relies on semi-quantitative information about the relative abundance of the analytes, keeping the gain accuracy for all of them along the analysis protocol requires very strict quality control procedures. Unfortunately, this is a point that has not received sufficient attention in most works.

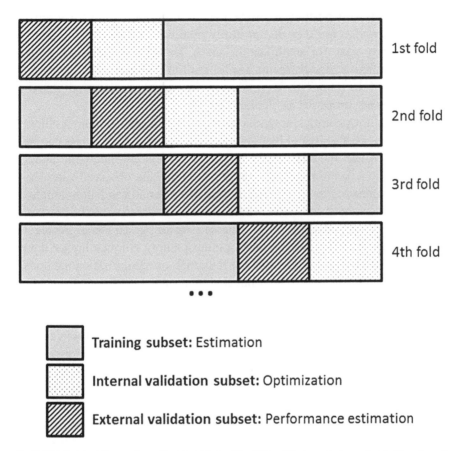

FIGURE 8.10   Illustration of a K-fold double CV validation strategy (with K = 4 and K = 5 for internal and external CV, respectively). The whole data set is divided into three separated subsets: (i) training subset to build a data model, (ii) internal validation subset to optimize the parameters of the model, and (iii) external validation subset to evaluate performance and generalization ability of the optimized model (final calibration model). The data set partition so that every sample belongs to the three subsets.

Since poor statistical analysis has been named as a cause of irreproducible research, we have also described some statistical issues. There is much controversy about p-values and the "p-hacking" phenomenon. Herein, we advocate to avoid binary thinking and consider other possibilities such as the volcano plot or Bayesian methods. We strongly recommend the construction of classification models instead of hypothesis testing, since predictive ability is not guaranteed even with a significant p-value.

Therefore, it is important to be aware that small p-values indicate population discrimination but they do not imply predictive power. Furthermore, p-values fairly vary for small sample sizes. Beyond hypothesis testing, building predictive models seems a safer path to understand the potential clinical use of the data. This is particularly important if relevant discriminant features are proposed as biomarkers.

Moreover, the potential presence of confounders is an important problem since they can be a source of bias. While some confounding factors can be controlled through metadata, there might be an uncontrolled confounder that is unknown to the research team and explains the observed variance. Our recommendation to fight this effect is to use external validation samples of a different origin or condition, where the confounder may not be present.

Finally, we have discussed the issue of overfitting. This is a prevalent problem since in omics research data dimensionality is much higher than the sample count. Thus, dimensionality reduction or regularization methods are a must. Overfitting has to be controlled by proper validation techniques, adjusting the model complexity to the available data. However, recent reviews on metabolomics studies have shown that sometimes validation is limited to internal validation. However, it is known that internal validation can lead to overoptimistic results. External validation is the preferred procedure to obtain an unbiased assessment on model performance, and particularly double CV becomes useful for small sample size. Furthermore, we would like to highlight that many predictive models are overfitted to the conditions of the study. The only way to check the limits of the model validity domain is by testing samples that are obtained under different conditions.

KEYWORDS

- **analytical robustness**
- **confounders**
- **metabolomics**
- **overfitting**
- **pitfals**
- **predictive models**
- **validation**

REFERENCES

Abaffy, T., Möller, M. G., Riemer, D. D., Milikowski, M., & DeFazio, R. A., (2013). Comparative analysis of volatile metabolomics signals from melanoma and benign skin: a pilot study. *Metabolomics, 9*, 998–1008.

Aggio, R. B., Costello, B. L., White, P., Khalid, T., Ratcliffe, N. M., Persad, R., & Probert, C. S., (2016). The use of a gas chromatography-sensor system combined with advanced statistical methods, towards the diagnosis of urological malignancies. *J. Breath Res., 10*, 017106.

Agilent, (2012). Metabolomics workflow – Discovery workflow guide, Application note 5990–7670EN, Agilent Technologies: Santa Clara, CA.

Altomare, D. F., Lena, M. D., Porcelli, F., Trizio, T., Travaglio, E., & Tutino, M., (2013). Exhaled volatile organic compounds identify patients with colorectal cancer. *Br. J. Surg., 100*, 144–150.

Amal, H., Leja, M., Funka, K., Skapars, R., & Sivins, A., (2016). Detection of precancerous gastric lesions and gastric cancer through exhaled breath. *Gut, 65*, 400–407.

Artursson, T., Eklov, T., Lundstrom, I., Martensson, P., Sjostrom, M., & Holmberg, M., (2000). Drift correction for gas sensors using multivariate methods. *J. Chemom., 14*, 711–723.

Azuaje, F., (2010). *Bioinformatics and Biomarker Discovery: Data Analysis for Personalized Medicine, 1st edn.*, John Wiley and Sons, Singapore, ISBN-10: 047074460X, pp. 248.

Bajtarevic, A., Ager, C., Pienz, M., Klieber, M., Schwarz, K., Ligor, M., & Ligor, T., (2009). Noninvasive detection of lung cancer by analysis of exhaled breath. *BMC Cancer, 9*, 348.

Baldi, P., & Brunak, S., (2001). *Bioinformatics: The Machine Learning Approach, 2nd ed.*, The MIT Press, Cambridge, MA, ISBN 0-262-02506-X.

Benjamini, Y., & Hochberg, Y., (1995). Controlling the false discovery rate: A practical and powerful approach to multiple testing. *J. Royal Stat. Soc., 57*, 289–300.

Berkel, J. V., Dallinga, J. W., Möller, G. M., Godschalk, R. W., & Moonen, E. J., (2010). A profile of volatile organic compounds in breath discriminates COPD patients from controls. *Respir. Med., 104*, 557–563.

Bickel, D., (2013). Simple estimators of false discovery rates given as few as one or two *p*-values without strong parametric assumptions. *Stat. Appl. Genet. Mol. Biol., 12*(4), 529–543.

Bijlsma, S., Bobeldijk, I., Verheij, E. R., Ramaker, R., Kochhar, S., Macdonald, I. A., Van Ommen, B., & Smilde, A. K., (2006). Large-scale human metabolomics studies: a strategy for data (pre-) processing and validation. *Anal. Chem., 78*(2), 567–574.

Bishop, C. M., (2006). *Pattern Recognition and Machine Learning*. Springer: Singapore.

Bland, J., & Altman, D. G., (1995). Multiple significance tests: the Bonferroni method. *Br. Med. J., 310*, 170.

Boulesteix, A., (2009). Over-optimism in bioinformatics. *Bioinformatics, 26*, 437–439.

Braga-Neto, U. M., & Dougherty, E. R., (2004). Is cross-validation valid for small-sample microarray classification? *Bioinformatics, 20*, 374–380.

Brereton, R. G., & Lloyd, G. R., (2014). Partial least squares discriminant analysis: taking the magic away. *J. Chemom., 28*, 213–255.

Brereton, R. G., (2006). Consequences of sample size, variable selection, and model validation and optimisation, for predicting classification ability from analytical data. *Trends Analyt. Chem.*, *25*(11), 1103–1111.

Broadhurst, D. I., & Kell, D. B., (2006). Statistical strategies for avoiding false discoveries in metabolomics and related experiments. *Metabolomics*, *2*(4), 171–196.

Broza, Y. Y., & Haick, H., (2014). Combined volatolomics for monitoring of human body chemistry. *Sci. Rep.*, *4*, 4611.

Bussche, J. V., Marzorati, M., Laukens, D., & Vanhaecke, L., (2015). Validated high resolution mass spectrometry-based approach for metabolomic fingerprinting of the human gut phenotype. *Anal. Chem.*, *87*, 10927–10934.

Chan, E. C., Pasikanti, K. K., & Nicholson, J. K., (2011). Global urinary metabolic profiling procedures using gas chromatography-mass spectrometry. *Nat. Protoc.*, *6*(10), 1483–1499.

D'Amico, A., Di Natale, C., Paolesse, R., Macagnano, A., Martinelli, E., Pennazza, G., Santonico, M., Bernabei, M., Roscioni, C., Galluccio, G., Bono, R., Agrò, E. F., & Rullo, S., (2008). Olfactory systems for medical applications. *Sens. Actuator. B-Chem.*, *130*(1), 458–465.

Dettmer, K., Aronov, P., & Hammock, B., (2007). Mass-spectrometry in metabolomics. *Mass Spectrom. Rev.*, *26*, 51–78.

Ducker, E., & Krapfenbauer, K., (2013). Pitfalls and limitations in translation from biomarker discovery to clinical utility in predictive and personalised medicine. *EPMA J.*, *4*(7).

Duda, R. O., Hart, P. E., & Stork, D. G., (2000). *Pattern Classification, 2nd edn.* John Wiley and Sons: New York, ISBN: 978-0-471-05669-0, pp. 680.

Dunn, W. B., Wilson, I. D., Nicholls, A. W., & Broadhurst, D., (2012). The importance of experimental design and QC samples in large-scale and MS-driven untargeted metabolomic studies of humans. *Bioanalysis*, *4*(18), 2249–2264.

Efron, B., & Gong, G., (1983). A leisurely look at the bootstrap, the jackknife, and cross-validation. *Am. Stat.*, *37*(1), 36–48.

Efron, B., & Tibshirani, R., (1997). Improvements on cross-validation: The 632+ Bootstrap Method. *J. Am. Stat. Assoc.*, *92*, 548–560.

Efron, B., (1983). Estimating the error rate of a prediction rule: improvement on cross-valdiation. *J. Am. Stat. Assoc.*, *78*, 316–331.

Efron, B., (2009). Empirical bayes estimates for large-scale prediction problems. *J. Am. Stat. Assoc.*, *104*(487), 1015–1028.

Efron, B., Tibshirani, R., Storey, J. D., & Tusher, V., (2001). Empirical bayes analysis of a microarray experiment. *J. Am. Stat. Assoc.*, *96*, 1151–1160.

Fernández-Albert, F., Llorach, R., García-Aloy, M., Ziyatdinov, A., Andrés-Lacueva, C., & Perera, A., (2014). Intesity drift removal in LC-MS metabolomics by common variance compensation. *Bioinformatics*, *30*, 2899–2905.

Fernie, A. R., Trethewey, R. N., & Krotzky, A. J., (2004). Metabolite profiling: from diagnostics to systems biology. *Nat. Rev. Mol. Cell. Biol.*, *5*(9), 763–769.

Filzmoser, P., Liebmann, B., & Varmuza, K., (2009). Repeated double cross validation. *J. Chemom.*, *23*, 160–171.

Flury, B., (1984). Common principal components in k groups. *J. Am. Stat. Assoc.*, *79*(388), 892–898.

Gardner, J. W., Shin, H. W., & Hines, E. L., (2000). An electronic nose system to diagnose illness. *Sens. Actuator. B-Chem.*, *70*, 19–24.

Gardner, M. J., & Altman, D. G., (1986). Confidence intervals rather than P values: estimation rather than hypothesistesting. *Br. Med. J.*, *292*, 746–750.

Gebregiworgis, T., & Powers, R., (2012). Application of NMR metabolomics to search for human disease biomarkers. *Comb. Chem. High Throughput Screen*, *15*(8), 595–610.

Ghosh, D., & Poisson, L. M., (2009). "Omics" data and levels of evidence for biomarker discovery. *Genomics*, *93*, 13–16.

Goodman, S. N., (1999a). The bayes factor. *Ann. Intern. Med.*, *130*(12), 1005–1013.

Goodman, S. N., (1999b). The P value fallacy. *Ann. Intern. Med.*, *130*(12), 995–1004.

Gosho, M., Nagashima, K., & Sato, Y., (2012). Study designs and statistical analyses for biomarker research. *Sensors*, *12*(7), 8966–8986.

Greenland, S., & Morgenstern, H., (2001). Confounding in health research. *Annu. Rev. Public Health*, *22*, 189–212.

Halsey, L. G., Curran-Everett, D., Vowler, S. L., & Drummond, G. B., (2015). The fickle P value generates irreproducible results. *Nat. Methods.*, *12*, 179–185.

Hayden, E. C., (2013). Weak statistical standards implicated in scientific irreproducibility. Nature News, November 11, 2013.

Ioannidis, J. P., (2005). Why most published research findings are false. *PLoS Med.*, *2*(8), e124.

Jepsen, P., Johnsen, S. P., Gillman, M. W., & Sorensen, H. T., (2004). Interpretation of observational studies. *Heart*, *90*, 956–960.

Kamleh, M. A., Ebbels, T. M., Spagou, K., Masson, P., & Want, E. J., (2012). Optimizing the use of quality control samples for signal drift correction in large-scale urine metabolic profiling studies. *Anal. Chem.*, *84*(6), 2670–2677.

Krilaviciute, (2015). Detection of cancer through exhaled breath: a systematic review. *Oncotarget*, *6*(36), 38643–38657.

Kukull, W. A., & Ganguli, M., (2012). Generalizability: The trees, the forest, and the low-hanging fruit. *Nature*, 1886–1891.

Kulasingam, V., & Diamandis, E., (2008). Strategies for discovering novel cancer biomarkers through utilization of emerging technologies. *Nature*, 588–299.

Lämmerhofer, M., & Weckwerth, W., (2013). *Metabolomics in Practice: Successful Strategies to Generate and Analyze Metabolic Data, 1st edn.* Wiley-VCH Verlag: Singapore.

Larrañaga, P., Calvo, B., Santana, R., & Bielza, C., (2006). Machine learning in bioinformatics. *Brief. Bioinform*, *7*, 86–112.

Leek, J. T., & Peng, R. D., (2015). *P*-values are just the tip of the iceberg. *Nature*, *520*, 612.

Machado, R. F., & Laskowski, D., (2005). Detection of lung cancer by sensor array analyses of exhaled breath. *Am. J. Respir. Crit. Care Med.*, *171*(11), 1286–1291.

Marco, S., & Gutierrez-Galvez, A., (2012). Signal and data processing for machine olfaction and chemical sensing: a review. *IEEE Sens. J.*, *12*(11), 3189–3214.

Marco, S., (2014). The need for external validation in machine olfaction: emphasis on health-related applications. *Anal. Bioanal. Chem.*, *406*, 3941–3956.

McLerran, D., Grizzle, W. E., Feng, Z., Bigbee, W. L., Banez, L. L., Cazares, L. H., Chan, D. W., Diaz, J., Izbicka, E., Kagan, J., Malehorn, D. E., Malik, G., Oelschlager, D., Partin, A., Randolph, T., Rosenzweig, N., Srivastava, S., Srivastava, S., Thompson, I. M., Thornquist, M., Troyer, D., Yasui, Y., Zhang, Z., Zhu, L., & Semmes, O. J., (2008). Analytical validation of serum proteomic profiling for diagnosis of prostate cancer: Sources of sample bias. *Clin. Chem.*, *54*, 44–52.

Miekisch, W., Herbig, J., & Schubert, J. K., (2012). Data interpretation in breath biomarker research: pitfalls and directions. *J. Breath Res.*, *6*(3), 036007.

Molinaro, A. M., Simon, R., & Pfeiffer, R. M., (2005). Prediction error estimation: a comparison of resampling methods. *Bioinformatics*, *21*, 3301–3307.

Monteiro, M., Carvalho, M., Bastos, M., & Pinho, P. G., (2013). Metabolomics analysis for biomarker discovery: advances and challenges. *Curr. Med. Chem.*, *20*, 257–271.

Nagle, H. T., Gutierrez-Osuna, R., & Schiffman, S. S., (1998). The how and why of electronic noses. *IEEE Spectrum*, *35*(9), 22–31.

Nuzzo, R., (2014). Statistical errors. *Nature*, *506*, 150–152.

Ojala, M., & Garriga, G. C., (2010). Permutations tests for studying classifier performance. *J. Mach. Learn. Res.*, *11*, 1833–1863.

Oliver, S. G., Winson, M. K., Kell, D. B., & Baganz, F., (1998). Systematic functional analysis of the yeast genome. *Trends Biotechnol.*, *16*(9), 373–378.

Padilla, M., & Bickel, D. R., (2012). Estimators of the local false discovery rate designed for small numbers of tests. *Stat. Appl. Genet. Mol. Biol.*, *11*(5), 1–42.

Padilla, M., Fonollosa, J., & Marco, S., (2013). Improving the Robustness of Odor Sensing Systems by Multivariate Signal Processing. In: *Human Olfactory Displays and Interfaces: Odor Sensing and Presentation*, Nakamoto, T., ed., IGI Global: Hershey, pp. 296–316.

Padilla, M., Perera, A., Montoliu, I., Chaudry, A., Persaud, K., & Marco, S., (2010). Drift compensation of gas sensor array data by orthogonal signal correction. *Chemom. Intell. Lab. Syst.*, *100*(1), 28–35.

Patti, G. J., Tautenhahn, R., Rinehart, D., Cho, K., Shriver, L. P., Manchester, M., Nikolskiy, I, Johnson, C. H., Mahieu, N. G., & Siuzdak, G., (2013). A view from above: Cloud plots to visualize global metabolomic data. *Anal. Chem.*, *85*(2), 798–804.

Patti, G. J., Yanes, O., & Siuzdak, G., (2012). Metabolomics: the apogee of the omics trilogy. *Nat. Rev. Mol. Cell. Biol.*, *13*, 263–269.

Petricoin, E. F., Ardekani, A. M., Hitt, B. A., Fusaro, V. A., Steinberg, S. M., Mills, G. B., Simone, C., Fishman, D. A., Kohn, E. C., & Liotta, L. A., (2002). Use of proteomic patterns in serum to identify ovarian cancer. *Lancet.*, *359*(9306), 572–577.

Pourhoseingholi, M. A., Baghestani, A. R., & Vahedi, M., (2012). How to control confounding effects by statistical analysis. *Gastroenterol. Hepatol. Bed Bench.*, *5*, 79–83.

Qiao, Z., Zhou, L., & Huan, J. Z., (2008). Effective linear discriminant analysis for high dimensional, low sample size data. *Proceedings of the World Congress on Engineering VOL II*, London, U. K., Ao, S. I., Gelman, L., Hukins, D. W. L., Hunter, A., Korsunsky, A. M., eds., IAEng: London.

Ransohoff, D. F., (2005). Bias as a threat to the validity of cancer molecular-marker research. *Nat. Rev. Cancer.*, *5*, 142–149.

Romain, A. C., André, Ph., & Nicolas, J., (2002). Three years experiment with the same tin oxide sensor arrays for the identification of malodorous sources in the environment. *Sens. Actuator. B-Chem.*, *84*(2–3), 271–277.

Saerys, Y., Inza, I., & Larrañaga, P., (2007). A review of feature selection techniques in bioinformatics. *Bioinformatics*, *23*(19), 2507–2517.

Schmekel, B., Winquist, F., & Vikström, A., (2014). Analysis of breath samples for lung cancer survival. *Anal. Chim. Acta.*, *840*, 82–86.

Schmidt, K., & Podmore, I., (2015). Current challenges in volatile organic compounds analysis as potential biomarkers of cancer. *J. Biomark.*, 981458.

Smolinksa, A., Blanchet, L., Buydens, L. M., & Wijmenga, S. S., (2012). NMR and pattern recognition methods in metabolomics: From data acquisition to biomarker discovery: A review. *Anal. Chim. Acta.*, *750*, 82–97.

Steyerberg, E. W., Bleeker, S. E., Moll, H. A., Grobbee, D. E., & Moons, K. G., (2003). Internal and external validation of predictive models: a simulation study of bias and precision in small samples. *J. Clin. Epidemiol.*, *56*(5), 441–447.

Storey, J. D., (2003). The positive false discovery rates: a Bayesian interpretation and the q-value. *Ann. Stat.*, *31*(6), 2013–2035.

Strathmann, F. G., & Hoofnagle, A. N., (2011). Current and future applications of mass spectrometry to the clinical laboratory. *Am. J. Clin. Pathol.*, *136*(4), 609–616.

Szymanska, E., Saccenti, E., Smilde, A. K., & Westerhuis, J. A., (2011). Double-check: Validation of diagnostic statistics for PLS-DA models in metabolomics studies. *Metabolomics*, *8*, 3–16.

Tautenhahn, R., Cho, K., Uritboonthai, W., Zhu, Z., Patti, G. J., & Siuzdak, G., (2012). An accelerated workflow for untargeted metabolomics using the METLIN database. *Nat. Biotechnol.*, *30*(9), 826–828.

Varma, S., & Simon, R., (2006). Bias in error estimation when using cross-validation for model selection. *BMC Bioinformatics*, *7*, 91.

Vinaixa, M., Samino, S., Saez, I., Duran, J., Guinovart, J. J., & Yanes, O., (2012). A guideline to univariate statistical analysis for LC-MS-based untargeted metabolomics-derived data. *Metabolites*, *2*, 775–795.

Westerhuis, J. A., Hoefsloot, H. C. J., Berkenbos-Smit, S., Vis, D. J., Smilde, A. K., Van Velzen, E. J. J., Van Duijnhoven, J. P. M., & Van Dorsten, F. A., (2008). Assessment of PLSDA cross validation. Metabolomics, *4*(1), 81–89.

Westhoff, M., Litterst, P., Maddula, S., Bödeker, B., & Baumbach, J. I., (2011). Statistical and bioinformatical methods to differentiate chronic obstructive pulmonary disease (COPD) including lung cancer from healthy control by breath analysis using ion mobility spectrometry. *Int. J. Ion Mobil. Spectrum*, *14*(4), 139–149.

WHO, World Health Organization, (2007). Quality assurance of pharmaceuticals: a compendium of guidelines and related materials (vol. *2*), WHO Library Cataloguing-in-Publication Data.

Worley, B., & Powers, R., (2013). Multivariate analysis in metabolomics. *Curr. Metabolomics*, *1*, 92–107.

Xia, J., Broadhurst, D. I., Wilson, M., & Wishart, D. S., (2013). Translational biomarker discovery in clinical metabolomics: an introductory tutorial. *Metabolomics*, *9*, 280–299.

Xu, Q. S., Liang, Y. Z., & Du, Y. P., (2004). Monte Carlo cross-validation for selecting a model and estimating the prediction error in multivariate calibration. *J. Chemom.*, *18*(2), 112–120.

Zelena, E., Dunn, W. B., Broadhurst, D., Francis-McIntyre, S., Carroll, K. M., Begley, P., O'Hagan, S., Knowles, J. D., & Halsall, A., (2009). HUSERMET Consortium, Wilson, I. D., Kell, D. B. Development of a robust and repeatable UPLC-MS method for the long-term metabolomic study of human serum. *Anal. Chem.*, *81*, 1357–1364.

CHAPTER 9

INSIGHT INTO KNAPSACK METABOLITE ECOLOGY DATABASE: A COMPREHENSIVE SOURCE OF SPECIES: VOC-BIOLOGICAL ACTIVITY RELATIONSHIPS

AZIAN AZAMIMI ABDULLAH,[1,2] MD. ALTAF-UL-AMIN,[1] and SHIGEHIKO KANAYA[1]

[1]*Graduate School of Information Science, Nara Institute of Science and Technology (NAIST), 8916-5, Takayama, Ikoma, 630-0192 Nara, Japan*

[2]*School of Mechatronic Engineering, Universiti Malaysia Perlis (UniMAP), Pauh Putra Campus, 02600 Arau, Perlis, Malaysia, Tel: +81-743-72-5952, Fax: +81-743-72-5329, E-mail: skanaya@gtc.naist.jp*

ABSTRACT

Volatile organic compounds (VOCs) are small molecules with low molecular weight that exhibit high vapor pressure under ambient conditions. VOCs are produced naturally by living organisms and have important roles in chemical ecology and human health. In chemical ecology, VOCs can serve as signaling molecules or semiochemicals passing information both within and between organisms. VOCs are also important in the healthcare field as they have the potential to be used as non-invasive diagnostic tools for detection of various human diseases such as cancer and gastrointestinal illness. Information on volatile emission from various organisms is scattered in the literature until now. However, there is still no available database describing VOCs and their biological activities.

To attain this purpose, we have developed the KNApSAcK Metabolite Ecology Database, which contains the information on the relationships between VOCs, emitting organisms and their biological activities. In this chapter, VOC database development and chemometrics methods for analyzing the relationships between chemical structures of VOCs and their corresponding biological functions are discussed in detail.

9.1   INTRODUCTION

Metabolomics is the scientific study of quantification of low mass compounds profiles and analysis of chemical processes involving metabolites in a comprehensive fashion. In general, metabolites can be divided into two groups: primary and secondary metabolites. Primary metabolites are directly involved in the normal growth, development and reproduction. On the other hand, secondary metabolites are not directly involved in these processes, but usually have important ecological functions, such as inter- or intra-species communication, antifungal, antimicrobial activities and also as a defense against pests and pathogens. Secondary metabolites are often colored, fragrant, or flavorful compounds and largely fall into three classes of compounds: alkaloids, terpenoids and phenolics. Small proportions produced by these secondary metabolites are volatile organic compounds (VOCs) that play important roles in chemical ecology and healthcare. VOCs can be defined as small compounds ranging in between C5 to C20 carbon count with a molecular weight in the range of 50 to 200 Daltons. They comprise of a diverse chemical group of organic compounds with various biological functions and have high vapor pressures under ambient conditions. Their high vapor pressure results from a low boiling point, which causes large numbers of molecules to evaporate from the liquid or solid form of the compound and enter the surrounding air, a trait known as volatility. Living organisms including human, animals, microorganisms and plants produce VOCs naturally. The naturally produced VOCs play important roles in communication between plants and they also serve as signaling molecules by passing information between organisms. For human and other animals, VOCs are important as scents and flavor of food. Recently, there are quite an increased number of researchers in utilizing VOCs as a biomarker to identify various kinds of diseases. Here, we elaborate further details the importance of VOCs for living organisms specifically in chemical ecology, agriculture and human healthcare.

9.2 IMPORTANCE OF VOCs

9.2.1 CHEMICAL ECOLOGY

VOCs constitute only a small proportion of the total number of secondary metabolites produced by living organisms, however, because of their important roles in chemical ecology specifically in the biological interactions between organisms and ecosystems, revealing and analyzing the roles of these VOCs is essential for understanding the interdependence of organisms. The total amount of VOCs emitted globally to the atmosphere is estimated to exceed 1 Pg per year, and these VOCs include mainly plant-produced VOCs, isoprene, monoterpenes and other oxygenated carbon compounds, such as herbivore-induced volatiles and green leaf volatiles (Iijima, 2014). Many studies have been performed that showed the emission of VOCs from plants occur as significant cues, signals, or defense responses to wounding, herbivore infestation, pathogen infection, and pollination. The emitted VOCs are responsible for internal and external communication between plants and herbivores, pathogens, pollinators, and parasitoids. Plants emit VOCs from their roots, leaves, fruits and flowers and use these compounds internally as defensive and signaling systems to induce levels of systemic acquired resistance (SAR) to pests and diseases. Some VOCs, such as methyl jasmonate α-pinene, camphene, and 1,8-cineol may inhibit the growth of other plants. VOCs produced by plant organs such as fruits and flowers also can act as external signaling molecules or semiochemicals by attracting pollinators and seed dispersers (Delory et al., 2016). They also contribute to the attraction of pest insects and beneficial insect predators in tritrophic interactions. Apart from plants, VOCs also act as a major communication among insects and other arthropods. Female insects use specific VOCs as sex pheromones to attract mates. Insects also use VOCs to mark pathways between nest and food and for defense.

9.2.2 AGRICULTURE

Conventional agricultural industry relies on a wide use of chemical pesticides and fertilizers. However, increased demand for organic products shows that consumers prefer reduced chemical use. Therefore, a novel sustainable agriculture needs to be developed for crop protection and prevention from using harmful chemicals. VOCs emitted by bacteria and fungi might have the potential as an alternative to the use of chemical pesticides to protect plants from pests and pathogens (Kanchiswamy et al., 2015a). It is because VOCs released by some rhizobacteria can enhance plant growth as well as inhibit the growth of other microorganisms. For example, acetoin and 2,3-butanediol released by rhizobacteria were found to promote the growth of Arabidopsis thaliana seedlings (Kai et al., 2016). A number of frequently emitted VOCs such as hexanal and 2-E-hexenal show antifungal activity and have been developed as an alternative to synthetic chemicals (Ayseli and Ayseli, 2016). Chemical ecologists also consider microbial VOCs as potential signaling molecules or semiochemicals that function as attractants and repellents to insects and other invertebrates. Pheromone traps are VOC based equipment for controlling pests without using harmful pesticides. In this strategy, pest insects may be diverted away from high-value crops using attractants, while simultaneously being repelled from high-value crops with repellents. Furthermore, natural enemies of insect pests, which are predators and parasitoids, may be simultaneously attracted making the use of semiochemicals a much more viable integrated management strategy than broad-spectrum chemical insecticides. For agriculture scientists, microbial VOCs are seen as biocontrol agents to control various phytopathogens and as biofertilizers for plant growth promotion (Kanchiswamy et al., 2015b). These examples indicate that the VOCs might have a potential impact on crop welfare and sustainable agriculture.

9.2.3   HUMAN HEALTHCARE

Microbial volatiles are widely used as biomarkers to detect human diseases. This is because bacteria have a recognizable metabolism that produces bacteria-specific VOCs, which might be used for non-invasive diagnostic purposes. For example, an electronic nose has been used to determine the causative bacteria responsible for diabetic foot infection by recognizing its volatiles (Yusuf et al., 2015). Hundreds of volatiles are emitted through the human body in breath, blood, skin, and urine. These compounds reflect the different metabolic conditions of an individual. Therefore, differences between the volatile profiles of individual humans can be used as an indicator to evaluate and monitor "disease" or "health" status (Buchbauer et al., 2015). Lourenco et al. have presented a comprehensive review of breath analysis in disease diagnosis using volatile profiles (Lourenço and

Turner, 2014). Besides breath, fecal headspace VOCs also can be used to diagnose gastrointestinal illness (Chan et al., 2016). Schmidt and Podmore (2015) have reviewed the analysis of VOCs using SPME as potential biomarkers of cancer. These examples indicate that disease-specific VOCs have potential as diagnostic olfactory biomarkers of infectious diseases, metabolic diseases, genetic disorders and other kinds of diseases.

9.3   VOC DATABASE DEVELOPMENT

With the explosively growing data scale, the development of biological databases incorporating different species has become a very important theme in big data biology. To address this need, we have developed KNAp-SAcK Family Databases (DBs), which have been utilized in a number of studies in metabolomics. The main window of KNApSAcK Family Databases is shown in Figure 9.1. These databases previously have been used to understand the medicinal usage of plants based on traditional and modern knowledge (Wijaya et al., 2014). To facilitate a comprehensive understanding of the interactions between the metabolites of organisms and the chemical level contribution of metabolites to human health, a metabolite activity DB known as the KNApSAcK Metabolite Activity DB has been constructed (Nakamura et al., 2014) and a network-based approach has been proposed to analyze the relationships between 3D structure and biological activities of the metabolites (Ohtana et al., 2014).

Advancement in analytical methods such as gas chromatography-mass spectrometry (GC-MS), proton transfer reaction mass spectrometry (PTR-MS) and selected ion flow tube mass spectrometry (SIFT-MS) have provided an opportunity to identify the volatile metabolites of living organisms in research laboratories. These analytical approaches generate a large amount of data and require specialized mathematical, statistical and bioinformatics tools to analyze such data. Despite the advances in sampling and detection by these analytical methods, only a few databases have been developed to handle these large and complex datasets. There are some VOC databases, which can be accessed freely. However, their applicability is often limited by several elements. Most of these databases only focus on volatiles, which are emitted by certain living organisms and have limited applications. None of these databases provide information on biological activities of VOCs and species-species interaction based on volatiles. To meet this purpose, we have developed a VOC database of microorganisms, fungi, and plants as well as human being, which comprises the relation between emitting species, volatiles and their biological activities (Abdullah et al., 2015). We have deposited the VOC data into KNApSAcK Metabolite Ecology Database, and this database is currently available at http://kanaya.naist.jp/MetaboliteEcology/top.jsp. Also, the database can be accessed online by clicking the corresponding button in the main window (Figure 9.1). Apart from the database development, we also analyzed the VOC data using hierarchical clustering and network clustering based on DPClus algorithm. In addition, we also performed the heatmap clustering based on Tanimoto coefficient as the similarity index between chemical structures to cluster all VOCs emitted by various biological species to understand the relationships between chemical structures of VOCs and their biological activities.

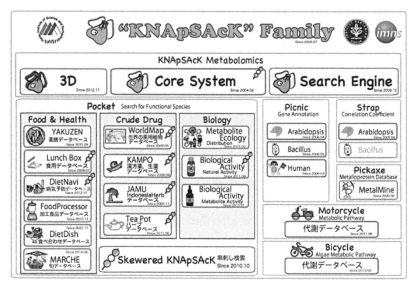

FIGURE 9.1   (See color insert.) The main window of the KNApSAcK Family Databases (http://kanaya.naist.jp/KNApSAcK_Family/).

The data were collected by an extensive literature search on PubMed (http://www.ncbi.nlm.nih.gov/pubmed) and Google Scholar (http://scholar.google.co.jp/). The information on VOCs, emitting species, target species and their biological activities were extracted and deposited into KNAp-SAcK Metabolite Ecology Database. The KNApSAcK Metabolite Ecology is also linked to the KNApSAcK Core and KNApSAcK Metabolite Activity database to provide further information on the metabolites and their biological activities. Data were divided into two types: (i) microorganisms species – VOC binary relations, and (ii) emitting species – VOC – target species triplet relations. At present, we have accumulated 1088 VOCs emitted by 517 microorganisms species and 341 VOCs emitted by other biological species including plants, fungi, animals and human. These VOC data have been deposited into KNApSAcK Metabolite Ecology Database, which allows users to search information on VOCs using the KNApSAcK compound ID and metabolite name. Figure 9.2 shows the main window of the KNApSAcK Metabolite Ecology Database, which shows the search types and search conditions. For search type, users can choose either partial or exact string matching searches by clicking the corresponding button, i.e., Partial or Exact (Figure 9.2A). Other check boxes can also be selected to specify different search conditions (Figure 9.2B) such as KNApSAcK compound ID (C_ID), metabolite name, species name and ecological category or localization. To search VOC data, users can input 'VOC' in the text box for the Ecological category/Localization category, select the corresponding checkbox and then click the List button (Figure 9.2C).

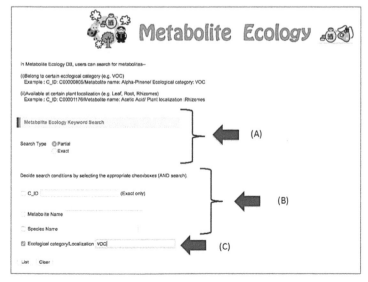

FIGURE 9.2 The main window of the KNApSAcK Metabolite Ecology Database (From Abdullah, et al. (2015). *Biomed Res. Int.*, 139–254. https://creativecommons.org/licenses/by/3.0/).

Part of the result retrieved by entering 'VOC' in the text box is shown in Figure 9.3. The attributes in the list are C_ID, which corresponds to the KNApSAcK compound identification (ID), metabolite name, species name (VOCs emitting species), ecological category/localization (VOC) and references (the source of the VOC's information), from left to right. During the literature search, it turned out that many VOCs do not have a KNApSAcK compound ID, but those might be biologically relevant. Therefore, those VOCs were also included in the database. For example, in the first line, the VOC with the name (+)-2-Carene (Figure 9.3A) does not have a KNApSAcK compound ID, though it was produced by *Solanum lycopersicum*. In future, we will find more information on these VOCs and assign the KNApSAcK compound ID to these metabolites. On the other hand, information related to the VOCs that have KNApSAcK compound ID can be obtained by clicking the C_ID as in Figure 9.3B.

INPUT WORD = [Match Type : Partial , Ecological category/Localization : VOC]

C_ID	Metabolite Name	Species Name	Ecological category/ Localization	Reference
-	(+)-2-Carene	Solanum lycopersicum	VOC	J Chem Ecol (2012) 38:1376_1388
-	(+)-4-Carene	Human (Urine)	VOC	Talanta 89 (2012) 360_368 (A)
-	(+)-Limonene	Solanum lycopersicum	VOC	J Chem Ecol (2012) 38:1376_1388
-	(+)-Sativene	Cladosporium cladosporioides	VOC	Sensors 2013, 13, 13969-13981
-	(+)-Sativene	Cochliobolus sativus	VOC	Fiers M, Lognay G, Fauconnier M-L, Jijakli MH (2013) PLoS ONE 8(6): e66805
-	(-)-Caryophyllene oxide	Solanum lycopersicum	VOC	J Chem Ecol (2012) 38:1376_1388
-	(2E,6E)-2,6-Nonadienoic acid	Human (Skin)	VOC	Wound Rep Reg (2010) 18 391_400
-	(E)-2-Hexenal	Solanum lycopersicum	VOC	Zebelo et al. BMC Plant Biology 2014,14:140
-	(E)-alpha-Bergamotene	Polygonum minus	VOC	Molecules 2014, 19, 19220-19242
-	(Z)-Myrtanol	Polygonum minus	VOC	Molecules 2014, 19, 19220-19242
-	-(Z)-Myrtanol	Polygonum minus	VOC	Molecules 2014, 19, 19220-19242
-	1,2,3,4,4a,5,6,8a-octahydro-4a,8-dimethyl-2-(1-methylethenyl)-Naphthalene	Polygonum minus	VOC	Molecules 2014, 19, 19220-19242
-	1,2,4-Trimethylbenzene	Human (Urine)	VOC	Talanta 89 (2012) 360_368
-	1,2-Benzenedicarboxylic acid	Polygonum minus	VOC	Molecules 2014, 19, 19220-19242
C00000805	alpha-pinene (A)	Cladosporium cladosporioides	VOC	Sensors 2013, 13, 13969-13977 (B)
C00000806	(-)-beta-Pinene	Solanum lycopersicum	VOC	Zebelo et al. BMC Plant Biology 2014,14:140
C00000816	beta-Pinene (A)	Proteaceae	VOC	J Chem Ecol (2013) 39:438_446
C00000823	Limonene	Polygonum minus	VOC	Molecules 2014, 19, 19220-19242
C00000823	Limonene	Proteaceae	VOC	J Chem Ecol (2013) 39:438_446
C00000853	Myrcene (A)	Proteaceae	VOC	J Chem Ecol (2013) 39:438_446
C00000861	Terpinolene (A)	Solanum lycopersicum	VOC	J Chem Ecol (2012) 38:1376_1388
C00001176	Acetic acid (A)	Burkholderia tropica	VOC	Bioengineered 4:4, 236_243 July/August 2013
C00001189	3-Methylbutanoic acid (A)	Staphylococcus aureus	VOC	PLOS Pathogens, May 2013, Volume 9, Issue 5

FIGURE 9.3 The results retrieved for VOC's search in the KNApSAcK Metabolite Ecology Database (From Abdullah, et al. (2015). *Biomed Res. Int.*, 139–254. https://creativecommons.org/licenses/by/3.0/).

Figure 9.4 shows the search results obtained by clicking the C_ID, C00000805, which were retrieved from the KNApSAcK Core Database.

Users can retrieve further knowledge of this metabolite, such as molecular formula, molecular weight, CAS Registry Number (CAS RN), molecular formula, 3D structure and other species information, which also produce the corresponding metabolite.

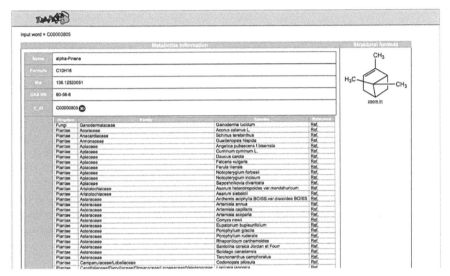

FIGURE 9.4 An example of the search results obtained by clicking the C_ID, C00000805, which were retrieved from the KNApSAcK Core Database (From Abdullah, et al. (2015). *Biomed Res. Int.,* 139–254. https://creativecommons.org/licenses/by/3.0/).

To understand the relationships between VOCs and their biological activities, we also integrate the KNApSAcK Metabolite Ecology Database with KNApSAcK Metabolite Activity Database. Information on biological activities of VOCs can be obtained by clicking the 'A' button in Figure Figure 9.3B. Figure 9.5 shows the search result of biological activities related to C_ID C00000805, which was retrieved from the KNApSAcK Metabolite Activity Database. The attributes in the list are C_ID, metabolite name, activity category, biological activity, target species and references, from left to right. Here, the metabolite known as alpha-pinene (C_ID C00000805) has several biological activities such as antimicrobial, antioxidant, biomarker, defense, enhance plant growth, anticholinesterase, and antifungal.

Metabolite Activity

INPUT WORD = [Match Type : Exact , C_ID : C00000805]

C_ID	Metabolite Name	Activity Category	Biological Activity (Function)	Target Species	Reference
C00000805	alpha-Pinene	Antimicrobial	Antimicrobial activity towards the tested microorganisms	Bacillus cereus	Ahmad et al.,Molecules,19,(2014),19220-19242
C00000805	alpha-Pinene	Antimicrobial	Antimicrobial activity towards the tested microorganisms	Enterococcus faecalis	Ahmad et al.,Molecules,19,(2014),19220-19242
C00000805	alpha-Pinene	Antimicrobial	Antimicrobial activity towards the tested microorganisms	Methilin-resistant Staphylococcus aureus (MRSA)	Ahmad et al.,Molecules,19,(2014),19220-19242
C00000805	alpha-Pinene	Antimicrobial	Antimicrobial activity towards the tested microorganisms	Salmonella entiriditis	Ahmad et al.,Molecules,19,(2014),19220-19242
C00000805	alpha-Pinene	Antioxidant	Leaf and stem have the highest antioxidant activity		Ahmad et al.,Molecules,19,(2014),19220-19242
C00000805	alpha-Pinene	Biomarker	Respiratory diseases (Cystic Fibrosis)	Human (sputum)	Goeminne et al.,Respiratory Research,13,(2012),13,87
C00000805	alpha-Pinene	Biomarker	The Digestive System (Irritable Bowel Syndrome)		Ahmed et al.,PLoS ONE,8,(2013),e58204
C00000805	alpha-Pinene	Defense	Defend against biotic stressors such as insects and pathogens	Spodoptera exigua	Zebelo et al.,BMC Plant Biology,14,(2014),140
C00000805	alpha-Pinene	Defense	Indirect defense responses	Trialeurodes vaporariorum	López et al.,J Chem Ecol,38,(2012),1376_1386
C00000805	alpha-Pinene	Enhance plant growth	Improved the growth of tobacco seedlings in vitro	Tobacco seedling	Paul et al.,Sensors,13,(2013),13969-13977
C00000805	alpha-Pinene	Anticholinesterase	Acetylcholinesterase inhibitory activities		Ahmad et al.,Molecules,19,(2014),19220-19242
C00000805	alpha-Pinene	Antifungal	Fungal growth inhibition	Colletotrichm gloesporioides	Tenorio-Salgado et al.,Bioengineered,4,(2013),236_243
C00000805	alpha-Pinene	Antifungal	Fungal growth inhibition	Fusarium culmorum	Tenorio-Salgado et al.,Bioengineered,4,(2013),236_243
C00000805	alpha-Pinene	Antifungal	Fungal growth inhibition	Fusarium oxysporum	Tenorio-Salgado et al.,Bioengineered,4,(2013),236_243

FIGURE 9.5 An example of the search result of biological activity related to C_ID C00000805, which was retrieved from the KNApSAcK Metabolite Activity Database. (From Abdullah, et al. (2015). Biomed Res. Int., 139–254. https://creativecommons.org/licenses/by/3.0/).

9.4 CHEMOMETRICS AND MACHINE LEARNING METHODS

9.4.1 CLUSTERING OF SPECIES BASED ON VOC SIMILARITY

Clustering is a machine learning method, which is the task of grouping a set of objects into the same group (cluster) based on similarity or distance measures. In this study, we utilized hierarchical clustering and graph clustering methods for classifying the VOC emitting species. Both methods are discussed separately in the following subheadings.

9.4.1.1 HIERARCHICAL CLUSTERING

We used hierarchical agglomerative clustering method, which starts out by putting each observation into its own separate cluster. The result of clustering is usually represented by a dendrogram. In our case, we used a Species vs. VOC matrix. Let this matrix be called M and M_{ik}=1 if the species i is related to the kth VOC or otherwise $M_{ik}=0$. Hierarchical methods require a distance *matrix*, and hence we determined the Euclidean distances between species. Euclidean distance, d between species i and species j can be calculated as follows:

$$d_{i,j} = k = (\Sigma_{k=1}^{n}(M_{ik}-M_{jk})^2)^{1/2}$$
$$cp_{nk} = E_{nk}/(d_k \times N_k) \tag{9.1}$$

Here, n is the number of VOCs, and there are 1088 VOCs in our data. Based on Euclidean distance, we performed the Ward's hierarchical clustering analysis using R, an open-source programming language.

9.4.1.2   GRAPH CLUSTERING BASED ON DPCLUS

DPClus is a graph clustering software, which has been developed based on a graph-clustering algorithm that can extract densely connected nodes as a cluster. This algorithm can be applied to an undirected simple graph $G = (N, E)$ that consists of a finite set of nodes N and a finite set of edges E. Two important parameters are used in this algorithm: density d_k and cluster property cp_{nk}. Density d_k of any cluster k is the ratio of the number of edges present in the cluster ($|E|$) and the maximum possible number of edges in the cluster ($|E|_{max}$). The cluster property of node n with respect to cluster k is represented by:

$$cp_{nk} = E_{nk}d_k \times N_k \tag{9.2}$$

N_k is the number of nodes in cluster k. E_{nk} is the total number of edges between the node n and the nodes of cluster k. In this study, we applied the DPClus algorithm to identify certain groups of microorganism species, based on VOC similarity. A network was constructed where a node represents a microorganism species, and an edge indicates high VOC similarity between the corresponding species pair. We selected 5% of the organism pairs based on the lower Euclidean distance between them. We used the non-overlapping mode with the following DPClus settings: Cluster property cp_{nk} was set to 0.5, density value d_k was set to 0.6, and minimum cluster size was set to 2.

9.4.2   CLUSTERING OF VOCs BASED ON CHEMICAL STRUCTURE SIMILARITY

We also performed a classification of VOCs based on their chemical structure similarity. In order to determine the similarity between two chemical compounds, we used Tanimoto coefficient as a similarity measure. The Tanimoto coefficient is defined as following equation, which is the proportion of the features shared by two compounds divided by their union:

$$\text{Tanimoto}_{A,B} = AB/(A+B-AB) \tag{9.3}$$

The variable *AB* is the number of features (or on-bits in a binary fingerprint) common in both compounds, while *A* and *B* are the number of features that are related to individual compounds, respectively. The Tanimoto coefficient has a range from 0 to 1 with higher values indicating greater similarity than lower ones. Additionally, a Tanimoto coefficient value larger than 0.85 indicates that the compared compounds may have similar biological activity. For the purpose of calculating Tanimoto coefficient, it is obligatory to assign fingerprints to the compounds. ChemMine package in *R* was used to generate molecular fingerprints and calculation of Tanimoto coefficient. 2-D compound structures in the generic structure definition file (SDF) format were obtained from PubChem database (https://pubchem.ncbi.nlm.nih.gov) and then, were imported into ChemmineR package in one batch file. The molecular descriptors are calculated during the SDF import and stored in a searchable descriptor database as a list object. Based on Tanimoto similarity measure between chemical structures, heatmap clustering was performed for classifying the VOCs. We also determined the *p*-values of the clusters based on hypergeometric distribution using following equation:

$$P\ value = 1 - \sum_{i=0}^{K-1} \frac{\binom{V}{i}\binom{N-V}{C-i}}{\binom{N}{C}} \tag{9.4}$$

Here *N* is the total number of VOCs; *C* is the size of a cluster and *V* and *K*, respectively, are the number of VOCs of a certain category in the whole data and in the cluster. The hypergeometric distribution is used to calculate the statistical significance of having drawn a specific K successes (out of C total draws) from the whole population. The test is often used to identify which subpopulations are over- or under-represented in a cluster. The calculated p-value implies the probability of getting K or more VOCs of a particular category in a cluster when the cluster is formed by random selection. Lower p-value indicates that the statistical significance is high. Our purpose is to relate a structure group to a biological activity if and only if the structure group is overrepresented by VOCs associated with that biological activity.

9.5 MINING OF KNAPSACK METABOLITE ECOLOGY DATABASE

9.5.1 CLUSTERING OF MICROORGANISMS-BASED ON VOC SIMILARITY

As disucussed in Section 9.3, the accumulated VOC data were divided into two types: (i) microorganisms species – VOC binary relations, and (ii) emitting species – VOC – target species triplet relations. This section focuses the clustering analysis result of the first type of data, which is the relationship between microorganism species and their emitting VOCs. Until now, we have accumulated 1088 compounds produced by 517 microorganisms. Figure 9.6 shows the log-log relation between the number of VOCs, M and the frequency of species, N. It shows that there are 92 species that emit only one type of VOC (Point x). Highest 50 types of VOCs are emitted by an individual species, and there are 14 such species in our present data (Point y).

From this statistical analysis, we can say that most microorganism species emit a few VOCs, which can act as their odor fingerprint. The information of emitting species and compounds has been converted into a 517×1088 binary matrix ("1" indicates presence while "0" indicates absence). The binary matrix then was used to calculate the Euclidean distance between species. From the Euclidean distance, hierarchical clustering of species was performed. The hierarchical dendrogram plot of microorganism species can be found in (Abdullah et al., 2015), where the dendrogram tree was cut into 50 clusters at a threshold height of 7. Out of 50 clusters, 3 clusters are consisting of only pathogenic bacteria; the other 3 are consisting of both pathogenic and non-pathogenic bacteria, and the rest 44 clusters are consisting of only non-pathogenic bacteria. These results imply that VOCs emitted by some pathogenic bacteria are different from those emitted by non-pathogenic bacteria. These results show consistency between VOC and pathogenicity-based classification of microorganisms.

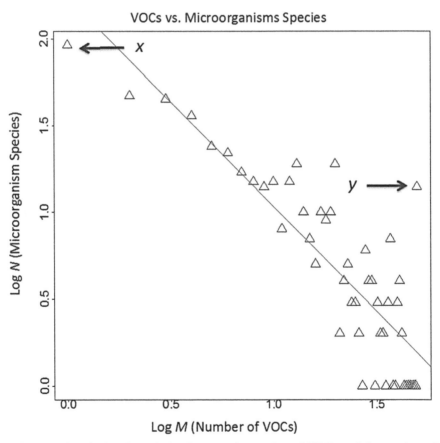

FIGURE 9.6   The log–log relation between the number of VOCs and the number of related microorganisms' species (From Abdullah, et al. (2015). *Biomed Res. Int.,* 139–254. https://creativecommons.org/licenses/by/3.0/).

In order to extract different and more information, we constructed a network by inserting edges between species for which the Euclidean distance is less than a threshold. The threshold was decided to include the lowest 5% distances as edges in the network. We then determined the high-density clusters in that network by applying the graph-clustering algorithm DPClus. Figure 9.7(a) shows the connected hierarchical graph of the clustering result while Figure 9.7(b) shows the independent nodes of the hierarchical graph, which indicates that these clusters do not interact with other clusters. Here, the green nodes represent clusters of micro-organism species, and the red edges represent the interaction between clusters. The radius of a green node in the hierarchical graph in Figure 9.7 is proportional to the logarithm of the number of nodes in the cluster it represents. The width of a red edge in the hierarchical graph between a pair of clusters is proportional to the number of edges between those clusters in the original graph. Overall, DPClus generated 50 clusters. From those, 20 clusters are connected nodes to each other while the rest 30 clusters are independent nodes. Only cluster 1 contains both pathogenic

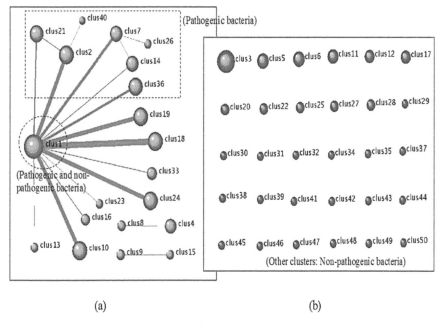

(a)   (b)

FIGURE 9.7   (See color insert.) Hierarchical graph of DPClus clustering result. (a) Connected nodes – nodes enclosed by dotted rectangle are consisting of only pathogenic bacteria, the only node enclosed by the dotted circle is consisting of both pathogenic and non-pathogenic bacteria and the rest nodes are consisting of only non-pathogenic bacteria. (b) Independent nodes – all nodes are consisting of non-pathogenic bacteria (Adapted from Abdullah, et al. (2015). *Biomed Res. Int.*, 139–254.).

and non-pathogenic microorganisms. Clusters 2, 7, 14, 21, 26 and 40 consist of only pathogenic bacteria while the other clusters are consisting of only non-pathogenic bacteria. These results imply that pathogenicity of microorganisms can be linked to characteristic combinations of identical VOCs emitted by them. Some of the pathogenic members of cluster 1 such as *Klebsiella pneumoniae, Escherichia coli, Staphylococcus aureus,* and *Pseudomonas aeruginosa* are very highly connected to other pathogenic clusters, for example, cluster 2 and 7. Figure 9.7(a) shows that cluster 2, 7, 14, 21, 26 and 40 are connected by red edges, which reflect VOC similarity between pathogenic microorganisms. Also, there is VOC based similarity between non-pathogenic species of cluster 1 and clusters 10, 13, 16, 18, 19, 23, 24, 33 and 36. The red edges between cluster 4 and 8 and between cluster 9 and 15 are also because of VOC similarity between non-pathogenic species of those clusters. Here it is noteworthy that the rest of the clusters consisting of non-pathogenic species are independent clusters, which implies that many non-pathogenic groups of species emit quite unique types of VOCs as shown in Figure 9.7(b). Here the internal nodes of a cluster are shown connected by green edges and its neighboring clusters are shown connected by red edges. To evaluate the stability of graph-clustering results by DPClus, we also clustered the networks generated by several random samplings of 80% or more edges of the original network. We found that DPClus can still cluster the microorganism species based on pathogenicity. The results of network clustering and hierarchical clustering are similar in the sense that both results indicated that VOC based classification of microorganisms is consistent with their classification based on pathogenicity. However, clustering by DPClus further revealed existence and non-existence of relations between different pathogenic and non-pathogenic groups of microorganisms. The classification achieved by DPClus is better in a sense it produced more clusters with 100% membership of either pathogenic or non-pathogenic microorganisms.

9.5.2   CLUSTERING OF VOCS-BASED ON STRUCTURAL SIMILARITY

The first types of data focused on microorganism species only but the second type of data include VOCs emitted by other biological species such as plants, animals, and humans. The second data that we have accumulated until now is 1044 species-species interactions via 341 VOCs associated with 11 groups of biological activities. The biological activities of VOCs are classified into two types: (i) chemical ecology-related activities, in which most VOCs involved in the interaction between species for the survival of organisms such as defense and antimicrobial; and (ii) human healthcare related activities, in which many VOCs are disease biomarkers and odors. From our accumulated data, 57.3% of the activities belong to chemical ecology such as antifungal, antimicrobial, attractant, defense, enhance plant growth, inhibit root growth and repellent activities and 42.7% are human health related activities such as disease biomarker, odor, anti-cholinesterase and antioxidant as shown in Figure 9.8.

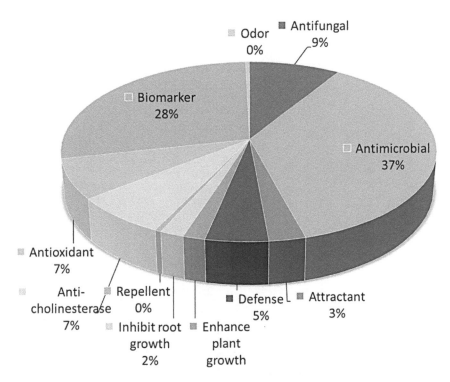

FIGURE 9.8   Pie chart showing the relative frequencies VOCs belonging to 11 biological activities (adapted from Abdullah, et al. (2015). *Biomed Res. Int.,* 139–254.).

There are many VOCs, which have several biological activities. Thus, it is important to investigate the relationships between VOCs and their biological activities statistically. Initially, we determined the pairwise chemical structural similarity between VOCs based on Tanimoto coefficient. 2-D compound structures in the generic structure definition file (SDF) format of all 341 VOCs were obtained from PubChem database and then, were imported into ChemmineR package in one batch file. We calculated the chemical structure similarity using Tanimoto coefficient. Then, we converted the Tanimoto similarity matrix into distance matrix by subtracting each of the similarity values from 1. Based on distance matrix, we performed heatmap clustering, and the result is shown in Figure 9.9. White and red colors indicate the extreme distance values of 0 and 1

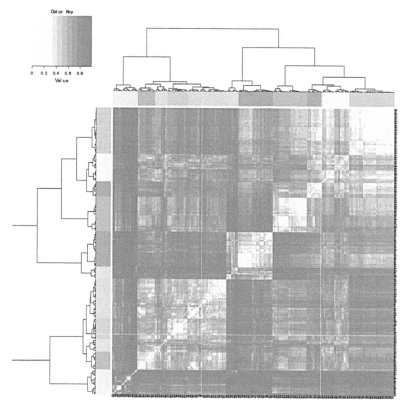

FIGURE 9.9 (See color insert.) Heatmap clustering of VOCs based on chemical structure similarity determined by Tanimoto coefficient (From Abdullah, et al. (2015). *Biomed Res. Int.*, 139–254. https://creativecommons.org/licenses/by/3.0/).

respectively and the intermediate distance values are indicated by the intensity of the red color. From the heatmap plot, we tentatively outlined 11 clusters of VOCs. The count of VOCs belonging to each activity group in each cluster is shown in Table 9.1. To assess the richness of VOCs of similar activity in individual clusters, we determined their p-values based on hypergeometric distribution, which are also shown in Table 9.1. The major types of chemical compounds belonging to each cluster and their corresponding biological activities are mentioned in Table 9.2. The detail description of chemical compounds in each cluster and their related biological activities can be found in (Abdullah et al., 2015). Additionally, we also compared several types of hierarchical clustering methods (single, complete, average, centroid, median linkage and Ward's method) with DPClus algorithm to cluster the chemical structures of VOCs using Tanimoto coefficient as a similarity measure. We found that Ward's method has the most matching clusters with DPClus while median has the least matching clusters (Abdullah et al., 2016). Compared to hierarchical clustering, DPClus can give a better visualization of how generated clusters interact with each other, and we found that VOCs belonging to the interacted clusters have similar chemical structure, which indicates possibilities of exhibiting similar biological activities. Both clustering methods show that there are strong links between the chemical structure of VOCs and their biological activities. Comparative activity relationships between chemical ecology and human health care activity will lead to the systematization of metabolomics combined with human and ecological metabolic pathways.

9.6   CONCLUSIONS

In this chapter, we have discussed a database of VOCs emitted by various living organisms including microorganisms, fungi, plants, animals and humans, which can be accessed at KNApSAcK Metabolite Ecology Database. Apart from VOC biological activities related to human healthcare, more than half of the biological activities are associated with chemical ecology. Hierarchical clustering and graph clustering by DPClus algorithm were utilized to extract specific clusters of microorganism species based on VOC similarity. We found consistency between VOC and pathogenicity based classification of microorganisms. Additionally, we also

TABLE 9.1 The Count of VOCs Belonging to Each Activity Group in Each Cluster and Their p-Value Based on Hypergeometric Distribution (From Abdullah, et al. (2015). Biomed Res. Int., 139–254 https://creativecommons.org/licenses/by/3.0/).

Biological Activity	Cluster ID (Count)	Cluster 1 (55)	Cluster 2 (33)	Cluster 3 (41)	Cluster 4 (18)	Cluster 5 (21)	Cluster 6 (25)	Cluster 7 (47)	Cluster 8 (15)	Cluster 9 (42)	Cluster 10 (14)	Cluster 11 (30)
Anti-cholinesterase	p-value	5.28×10^{-7}	4.849×10^{-4}	0.6181	0.9109	1.274×10^{-2}	0.9147	1	0.3596	0.9994	1	1
	(Count)	(26)	(15)	(8)	(2)	(9)	(3)	(0)	(4)	(2)	(0)	(0)
Antifungal	p-value	0.9128	0.9115	0.5399	2.561×10^{-2}	1	0.5176	0.6403	0.6571	0.3099	1	0.3291
	(Count)	(2)	(1)	(3)	(4)	(0)	(2)	(3)	(1)	(4)	(0)	(3)
Antimicrobial	p-value	2.10×10^{-6}	9.696×10^{-4}	0.6898	0.9281	1.871×10^{-2}	0.8246	0.9999	0.4049	0.9997	1	1
	(Count)	(26)	(15)	(8)	(2)	(9)	(4)	(1)	(4)	(2)	(0)	(0)
Antioxidant	p-value	5.28×10^{-7}	4.849×10^{-4}	0.6181	0.9109	1.274×10^{-2}	0.9147	1	0.3596	0.9994	1	1
	(Count)	(26)	(15)	(8)	(2)	(9)	(3)	(0)	(4)	(2)	(0)	(0)
Attractant	p-value	0.9708	0.8144	1	0.4831	1	0.1661	3.829×10^{-2}	0.1356	1.983×10^{-2}	1	1
	(Count)	(2)	(2)	(0)	(2)	(0)	(4)	(8)	(3)	(8)	(0)	(0)
Biomarker	p-value	1	0.9999	1.835×10^{-3}	0.7944	1.444×10^{-2}	0.9821	4.42×10^{-5}	0.9948	6.071×10^{-2}	1.036×10^{-2}	1.963×10^{-3}
	(Count)	(8)	(11)	(34)	(10)	(18)	(11)	(41)	(5)	(31)	(13)	(26)
Defense	p-value	3.35×10^{-9}	9.258×10^{-2}	0.9758	0.6764	0.7594	0.8418	0.9987	0.8668	0.9787	1	1
	(Count)	(22)	(7)	(2)	(2)	(2)	(2)	(1)	(1)	(2)	(0)	(0)
Enhance Plant Growth	p-value	6.01×10^{-2}	1	1	1	1	3.531×10^{-3}	0.7778	1	1	1	0.6069
	(Count)	(4)	(0)	(0)	(0)	(0)	(4)	(1)	(0)	(0)	(0)	(1)
Inhibit root growth	p-value	0.1749	0.8632	0.9183	1	0.7111	4.111×10^{-2}	0.7672	0.5847	0.7062	0.5591	0.8347
	(Count)	(5)	(1)	(1)	(0)	(1)	(4)	(2)	(1)	(2)	(1)	(1)
Odor	p-value	1	1	1	1	1	2.29×10^{-5}	1	1	1	1	1
	(Count)	(0)	(0)	(0)	(0)	(0)	(4)	(0)	(0)	(0)	(0)	(0)
Repellent	p-value	1	7.551×10^{-2}	1	1	1.871×10^{-3}	1	1	1	1	1	1
	(Count)	(0)	(2)	(0)	(0)	(3)	(0)	(0)	(0)	(0)	(0)	(0)

TABLE 9.2 Summary of Clustering Result and Its Descriptions Related to Chemical Structures and Biological Activities (From Abdullah, et al. (2015). Biomed Res. Int., 139–254 https://creativecommons.org/licenses/by/3.0/).

Cluster ID (Count)	Description on chemical structures	Related biological activities
Cluster 1 (55 VOCs)	All compounds are terpenoids. 15 VOCs are monoterpenoids (10 carbon) and 40 VOCs are sesquiterpenoids (15 carbon).	Anti-cholinesterase, antimicrobial, antioxidant and defense.
Cluster 2 (33 VOCs)	17 VOCs are alcohol, aldehyde, ketone, epoxide and ester of terpenoids. The other VOCs are alcohol, aldehyde, carboxylic acid, ester and ketone of straight-chain alkenes.	Anti-cholinesterase, antimicrobial and antioxidant.
Cluster 3 (41 VOCs)	Alkanes.	Biomarker.
Cluster 4 (18 VOCs)	Alkenes.	Antifungal.
Cluster 5 (21 VOCs)	Aldehyde, ester, carboxylic acid and ketone of C8-C18 alkanes.	Anti-cholinesterase, antimicrobial, antioxidant, biomarker and repellent.
Cluster 6 (25 VOCs)	21 VOCs are alcohol and ether of C3-C8 alkanes.	Enhance plant growth, inhibit root growth and odor.
Cluster 7 (47 VOCs)	45 VOCs are ester, carboxylic acid, ketone and aldehyde of non-cyclic C2-C9 alkanes.	Attractant and biomarker.
Cluster 8 (15 VOCs)	VOCs consist of epoxide, ethers, esters and alcohols.	-
Cluster 9 (42 VOCs)	24 VOCs are aromatic alcohols, carboxylic acids, esters, ketones, and ethers. 16 VOCs are aromatic compounds consisting of C and H atoms. One VOC consists of C, H and Br atoms. One VOC is an alkane ester.	Attractant.
Cluster 10 (14 VOCs)	Aromatic compounds. 12 VOCs are hetero-aromatic compounds that consist of one or more sulfur, nitrogen or oxygen atoms.	Biomarker.
Cluster 11 (30 VOCs)	VOCs are quite diverse in chemical elements, C0-C6 small molecules.	Biomarker.

compared several types of hierarchical clustering methods with DPClus clustering to classify VOCs using fingerprint-based similarity measure between chemical structures. Our research indicates that similar chemical structures of VOCs indicate possibilities of exhibiting similar biological activities. In future, more VOCs can be accumulated, and comprehensive analysis can be performed in the context of human healthcare and chemical ecology. The KNApSAcK Metabolite Ecology Database may be useful for the discovery of novel agricultural tools and also for the non-invasive identification of biomarkers in the medical diagnostic field as well as a systematic research in various omics fields, especially metabolomics integrated with ecosystems.

KEYWORDS

- **chemometrics**
- **data mining**
- **database**
- **ecology**
- **KNApSAcK**
- **machine learning**
- **metabolite**
- **VOCs**

REFERENCES

Abdullah, A. A., Altaf-Ul-Amin, M., Ono, N., Sato, T., Sugiura, T., Morita, A. H., Katsuragi, T., Muto, A., Nishioka, T., & Kanaya, S., (2015). Development and mining of a volatile organic compound database. *Biomed Res. Int.*, 139254.

Abdullah, A. A., Altaf-Ul-Amin, M., Ono, N., Yusuf, N., Zakaria, A., Nishioka, T., & Kanaya, S., (2016). Comparison of clustering methods in the context of chemical structure similarity based classification of VOCs. *Procedia Chem.*, *20*, 40–44.

Ayseli, M. T., & Ayseli, Y. İ., (2016). Flavors of the future: Health benefits of flavor precursors and volatile compounds in plant foods. *Trends Food Sci. Technol.*, *48*, 69–77.

Buchbauer, G., & Buljubasic, F., (2015). The scent of human diseases: A review on specific volatile organic compounds as diagnostic biomarkers. *Flavour Fragr. J.*, *30*, 5–25.

Chan, D. K., Leggett, C. L., & Wang, K. K., (2016). Diagnosing gastrointestinal illnesses using fecal headspace volatile organic compounds. *World J. Gastroenterol.*, *22*(4), 1639.

Delory, B. M., Delaplace, P., Fauconnier, M. L., & Du Jardin, P., (2016). Root-emitted volatile organic compounds: can they mediate belowground plant-plant interactions? *Plant Soil*, *402*(1–2), 1–26.

Iijima, Y., (2014). Recent advances in the application of metabolomics to studies of biogenic volatile organic compounds (BVOC) produced by plant. *Metabolites*, *4*, 699–721.

Kai, M., Effmert, U., & Piechulla, B., (2016). Bacterial-plant-interactions: approaches to unravel the biological function of bacterial volatiles in the rhizosphere. *Front. Microbiol.*, *7*, 108.

Kanchiswamy, C. N., Malnoy, M., & Maffei, M. E., (2015a). Bioprospecting bacterial and fungal volatiles for sustainable agriculture. *Trends Plant Sci.*, *20*(4), 206–11.

Kanchiswamy, C. N., Malnoy, M., & Maffei, M. E., (2015b). Chemical diversity of microbial volatiles and their potential for plant growth and productivity. *Front. Plant Sci.*, *6*, 151.

Lourenço, C., & Turner, C., (2014). Breath analysis in disease diagnosis: Methodological considerations and applications. *Metabolites*, *4*(2), 465–498.

Nakamura, Y., Afendi, F. M., Parvin, A. K., Ono, N., Tanaka, K., Morita, A. H., Sato, T., Sugiura, T., Altaf-Ul-Amin, M., & Kanaya, S., (2014). KNApSAcK metabolite activity database for retrieving the relationships between metabolites and biological activities. *Plant Cell Physiol.*, *55*(1), e7–e7.

Ohtana, Y., Abdullah, A. A., Altaf-Ul-Amin, M., Huang, M., Ono, N., Sato, T., Sugiura, T., Horai, H., Nakamura, Y., Lange, K. W., Kibinge, N. K., Katsuragi, T., Shirai, T., & Kanaya, S., (2014). Clustering of 3D-structure similarity based network of secondary metabolites reveals their relationships with biological activities. *Mol. Inform.*, *33*(11–12), 790–801.

Schmidt, K., & Podmore, I., (2015). Solid phase microextraction (SPME) method development in analysis of volatile organic compounds (VOCS) as potential biomarkers of cancer. *J. Mol. Biomark. Diagn.*, *6*, 6.

Wijaya, S. H., Husnawati, H., Afendi, F. M., Batubara, I., Darusman, L. K., Altaf-Ul-Amin, M., Sato, T, Ono, N., Sugiura, T., & Kanaya, S., (2014). Supervised clustering based on DPClusO: prediction of plant-disease relations using Jamu formulas of KNApSAcK database. *Biomed. Res. Int.*, 831751.

Yusuf, N., Zakaria, A., Omar, M. I., Shakaff, A. Y. M., Masnan, M. J., Kamarudin, L. M., Abdul Rahim, N., Zakaria, N. Z. I., Abdullah, A. A., Othman, A., & Yasin, M. S., (2015). *In-vitro* diagnosis of single and poly microbial species targeted for diabetic foot infection using e-nose technology. *BMC Bioinformatics*, *16*(1), 158.

INDEX

1

1, 8-cineol, 227
12 (13)-epoxyoctadecenoic acid (12(13)-EpOME), 108
12(13)-EpOME, 108
12, 13- dihydroxyoctadecenoic acid (12, 13-DiHOME), 108
12-hydroxyeicosatetraenoic acid (12-HETE), 108
1-octen-3-ol, 3-methylfurane, 128

2

2-butanol, 131, 133
2D models, 60
2-isopropryl 3-methoxypyrazine, 133
2-methyl-isorbenol, 133

3

3D cell culture systems, 116
3D structure, 229, 233
3-methylfuran, 128, 131, 133

8

8-isoprostane, 181

9

9, 12, 13- TriHOME, 108
9, 12, 13- trihydroxyoctadecenoic acid (9, 12, 13- TriHOME), 108

A

Acceleration energy, 42
Acetaldehyde, 111, 174, 182, 184
Acetic acid, 112
Acetoin, 228
Acetone, 64, 107, 163, 174, 179, 181, 182
Acid–base metabolism, 65
Acidic functional groups, 39

Acinetobacter, 137, 180
Activated charcoal pads, 134, 142
Acute respiratory distress syndrome (ARDS), 184
Adenosine triphosphate (ATP) synthesis, 57, 184
Adenosquamous carcinoma, 67
Aerobic glycolysis, 57
Agricultural rural emissions, 123, 126
Agriculture, 141, 226–228
    human healthcare, 226
    pathological and clinical diagnoses, 141
Air
    circulating, 94
    water partition coefficients, 161
Airborne microorganisms, 134
Airway
    inflammation, 54
        activity, 177
    metabolite analysis, 181
Alcohols, 33, 63, 112, 128, 139
Aldehydes, 63, 107, 112, 128, 139, 140, 144, 146, 181, 182
Algorithms, 80, 151, 177
Aliphatic alcohols, 111
Alkaloids, 226
Alkanes, 56, 107, 122, 128, 181, 182
Allergy testing, 182
Alpha-pinene, 233
Alternaria alternata, 138
Alveolar
    air, 161, 164
    breath, 163, 164, 166
    capillary barrier, 163
    gradients, 86
    portion, 164
Alzheimer's and Parkinson's diseases, 180
Ambient conditions, 225, 226
Amines, 33, 139, 140, 144, 146
Amino acid, 7, 111

Ammonia, 39, 176, 184
Amyloid beta (Aβ), 180
Anaerobic, 65, 106
Analysis
    of variance (ANOVA), 110
    platform, 163
    procedures, 70
Analytes, 35–37, 170, 172, 174, 175, 198,
    216
Analytical
    approaches, 123, 126, 229
    capabilities, 177
    chemistry, 98, 200
    column, 33, 35
    framework, 123
    instruments, 79
    methods, 19, 58, 169, 229
    precision, 162
    procedure, 202–204
    robustness, 202, 216, 218
    space, 37
    speciation analysis, 123, 126
    technique, 6, 33, 39, 40, 79, 125, 146,
        150, 151, 171, 175, 200
    tools, 79
Angiogenesis, 56
Anomalies, 18
Anthocyanins, 83
Anthropogenic, 123, 126, 133, 134, 149
    activities, 123, 126
    sources, 134
Anticholinesterase, 233
Anticoagulated, 18
Antifungal, 226, 228, 233, 241
Antimicrobial, 226, 233, 241
Antioxidant, 56, 233, 241
Antiviral treatments, 180
Apoptosis, 56
Aromatic
    aliphatic hydrocarbons, 63
    compounds, 108, 111, 128
Arthrobacter globiformis, 136
Artificial
    neural network, 80, 85
    olfaction
        analysis, 197
        systems, 79, 80–83, 89–91, 93, 97–100

Aspergillosis, 113
Aspergillus
    flavus, 133
    fumigatus, 113, 126, 133
    niger, 133
    terreus, 133
Asthma, 6, 7, 65, 85, 108, 124, 169, 177,
    182, 183
Asthmatics, 183
Australian Government Department of
    Health, 185
Automated mass spectral deconvolution
    and identification system (AMDIS), 45,
    149
Autophagy, 12, 182

B

Bacterial
    cell culture models, 110
    colonization, 110
    infection, 111
    populations, 93
Basement membrane, 57
Bayes Factor, 209
Benign
    conditions, 87–89
    larynx operation, 88
    neck operation, 88
    prostatic
        hyperplasia, 92, 211
        hypertrophy, 91
Benzaldehyde, 115
Benzenoids, 112
Big data biology, 229
Binary matrix, 237
Bioaerosol, 123–126, 141, 143, 150, 151
    characterization, 125, 150
    mass spectrometer (BMAS), 125
    monitoring methods, 124
    studies, 124, 125
Bioaerosols emissions, 124
Biochemical
    diversity, 109
    metabolical origin, 64
    pathways, 114, 198
Biocontrol agents, 228

Biofertilizers, 228
Biofluid, 9, 16, 168, 198
BioGuardian air sampler, 143
Bioinformatics, 14, 17, 200, 213, 229
Biological
  activities, 225, 226, 229–231, 233,
    241–243, 246
  condition groups, 200
  databases, 229
  fluid samples, 170
  interpretation, 171
  non-biological particles, 125
  origin, 112
  particles, 124, 125
  samples, 14, 45, 177, 198, 199
  signatures, 13
  species, 230, 231, 240
  substances, 58
  system, 6, 12–14, 160, 185
Biomarker, 7, 55, 56, 108, 109, 115, 116, 163,
  178, 185, 199–201, 203, 209, 226, 241
Biomedical applications, 3, 98
Biomolecular organization, 12
Biomolecules, 169
Biopsy, 84, 91, 115
BioSampler, 143
Biosensing, 125
Biosensor-based multisensorial system
  for mimicking Nose, Tongue, and
  Eyes(BIONOTE), 83, 84
Bio-waste
  facilities, 126
  processing plants, 124
Bladder cancer, 62, 91
Blastomycosis, 113
Blood
  chemistry, 18
  clots, 183, 184
  concentrations, 58
  gas interface, 57, 63
  pressure, 18, 183
  samples, 62, 63
  stream, 65, 163
  tests, 67
Body fluid, 5, 62–64, 79, 80, 82, 97–100,
  197–199
Bone scan, 67

Bonferroni correction, 206
Bowel cleansing, 93
Breast cancer (BC), 66, 89, 90, 96
Breath
  analysis, 6, 12, 53, 57, 58, 64–66,
    68–71, 82, 84, 85–90, 98, 106, 107,
    116, 160, 167, 172, 174–178, 181,
    182, 185, 199, 215, 228
  analyzers, 184
  biomarker panels change with physi-
    ology, 108
  collection, 70, 106, 164, 165, 169
  condensate, 168, 170–172
  diagnostic studies, 110
  fingerprinting, 71
  profiles, 164
  pulmonary diagnostics, 112
  samples, 54, 58, 69, 80, 81, 85, 87–90,
    97, 108, 112, 126, 160, 162, 169, 170,
    176, 179, 181, 182, 184, 185
  sampling considerations, 161
  testing techniques, 180
  volatile tests, 113
Breathing patterns, 164
Breathomics, 6, 160, 161, 166, 178,
  180–182, 185, 186
Breathomics research, 160
Breathprint, 90, 169, 177, 178, 180
Bronchial epithelial cells, 114
Bronchoalveolar lavage, 180
Burholdaria cepacia, 111
Butoxyethoxyethanol/2-nonen-1-ol, 113
Butyric acid, 111

C

Calcium metabolism disorders, 65
Calibration routines, 203
Cancer
  cell, 56, 57, 59–61, 63, 68, 94–96, 114,
    115
    clone, 59
    metabolism, 57
    population, 59
  management, 55–58, 62–64
  progression, 64
  related deaths, 55

screening, 66
stem cells (CSC), 61
therapies, 12
tissue, 63
Cancerous cells, 79, 96
Capillary electrophoresis (CE), 4, 173,
    174, 177
    capillary, 173
    sample, 173
Carbohydrates, 13
Carbon
    bonds, 57
    compounds, 227
    dioxide, 4, 39, 164, 176, 179
    monoxide, 176, 178
    nanotube, 125
Carbonaceous adsorbents, 144
Carboxylic acids, 146
CAS Registry Number (CAS RN), 233
Cell
    cloning, 59
    culture, 97, 106–108, 112, 115, 116,
        151, 181
    growth, 61
    membranes, 56
    metabolic activity assays, 67
    metabolism, 56, 61
    mixtures, 59
    treatments, 115
    types, 59, 107, 114–116
Cellular
    level, 89, 93, 179
    proliferation, 96
Cerumen, 9
Charcoal pads, 134, 139, 142, 146
Chemical
    analysis, 17, 33, 37, 39, 125
    characterization, 125, 151
    component, 34, 37
    compounds, 40, 160, 236, 243
    derivatization, 172
    ecology, 225–227, 241, 243, 246
    groups, 128
    identification, 15
    insecticides, 228
    instrumentation, 200
    ionization (CI), 38, 39, 147, 148

properties, 134
sensor arrays, 5
separation, 36
structural information, 39
structure, 39, 109, 128, 226, 230, 236,
    242, 243, 246
techniques, 124, 125, 127
volatility, 32, 33, 39
warfare agents, 175
Chemiresistors, 80
ChemMine package, 236
Chemometrics, 70, 148, 200, 226, 246
    machine learning methods, 234
    software, 150
Chemotaxonomy/chemotyping, 150
Chemotherapy, 59, 60, 66–68, 85
    resistance assays, 67
Chest X-ray, 67, 184
Chromatogram, 148, 149, 172, 173
Chromatograph, 171
Chromatographic
    alignment, 19
    profile, 36, 37
    separation, 5, 32, 34, 35, 37, 40, 146
    syringe, 91
Chromatography, 4, 8, 31–33, 38, 39, 54,
    108, 144, 169, 172, 175, 202
Chronic
    exposure, 124
    kidney disease (CKD), 169
    liver disease (CLD), 181
    obstructive pulmonary disease (COPD),
        11, 65, 83, 85, 169, 174, 177, 181,
        184, 199
Circulating tumor cells (CTC), 62
Classification models, 93, 97, 209, 217
Clinical
    application, 58, 62, 65, 67, 69, 71, 201
    assay development, 56
    chemistries, 18
    disease, 7, 56
    environment, 70, 185
    implementation, 56, 69–71
    model, 111
    practice, 13, 53–55, 56, 58, 69, 171
Clustering methods, 234, 243
Collision cell, 41, 43

Colonization, 111, 180
Colonoscopy, 93
Colorectal
  cancer (CC), 90, 92, 93, 181, 199
  carcinoma, 93
Colorimetric sensors, 80, 83
Commonwealth Scientific and Industrial
  Research Organization (CSIRO), 159, 185
Complexity optimization, 214
Computational
  task, 200
  techniques, 200
Computed tomography (CT), 67, 184
Configurational isomers, 174
Confounders, 70, 211, 212, 218
Confounding factors, 94, 202, 205, 210,
  212, 218
Congestive heart failure, 183
Conidiation, 112
Contamination, 4, 63, 69, 106, 128, 141,
  148, 162
Continuous positive airways pressure
  (CPAP), 164
Conventional agricultural industry, 227
Coordination of standards in metabolomics
  (COSMOS), 14
Crossvalidation (CV) methods, 162, 209,
  214–218
Cryofocusing, 176
Culinary applications, 112
Cyclone devices, 142
  glass impingers, 134, 142, 143
Cyranose, 82, 86, 93, 95, 96, 98
Cystic fibrosis, 110–112, 169, 184

D

Data
  acquisition parameters, 19
  analysis, 20, 53, 65, 71, 106, 148, 160,
    197, 200, 205, 206, 212, 214
  base, 16, 17, 20, 112, 149, 151, 225,
    230–232, 236, 242, 243, 246
  bases (DBs), 229
  development, 230
  mining, 246
  modeling techniques, 106

pre-processing, 19
processing, 15, 201, 202, 205, 216
visualization, 106
Deconvolution, 45, 109, 148
Degradation/oxidation, 69
Dendrogram tree, 237
Detection
  method, 106
  VOCs, 108
Diabetes, 65
Diagnosis of cancers (airways), 86
Diagnostic
  olfactory biomarkers, 229
  platform, 106
  technologies, 80
  tests, 86, 177
Dietary restrictions, 18
Differential mobility spectrometers
  (DMS), 175
Dimethyl
  benzaldehyde, 108
  disulfide, 128, 132
  sulfide, 115
  sulfoxide (DMSO), 181
Dipole–dipole interaction, 34
Direct
  analysis in real time (DART), 40
  current (DC), 41
  injection mass analysis techniques, 40
Discriminant
  analysis, 86, 91–94, 213
  function analysis, 80, 85, 87, 88, 90, 97
    algorithm, 87
Discrimination model, 85
Disease
  infections, 160
  mechanisms, 13
  process, 55
  progression, 55
DNA
  RNA techniques, 124
  sequencing analysis, 150
Dolphin breath, 107
DPClus algorithm, 230, 235, 243
Drug combinations, 67
Dual ionization source, 44

Dulbecco's Modified Eagle's Medium,
  95, 96
Dying/dead cells, 59

E

Early Detection Research Network
  (EDRN), 56
Echocardiogram, 184
Ecological
  category/localization, 231
  functions, 226
Ecology, 225, 241, 246
Effect size (ES), 206, 208
Electric field, 38, 41, 42, 44, 173, 175
Electrical conductivity, 80
Electrocardiogram (ECG), 184
Electrode, 41, 42, 44
Electrodynamics squeezing, 44
Electron impact ionization (EI), 38, 147,
  148, 172
Electronic
  collections, 15
  nose, 79, 88, 94, 95, 96, 99, 141, 151,
    170, 180, 228
    applications, 7
Electrophoresis, 173
Electrospray ionization, 40
Electrostatic fields, 44
Elucidation, 39, 40, 110
Emphysema, 65
Endogenous
  exogenous, 8, 12
    chemicals, 107
  products, 12
Endoscopic examination, 87
ENose
  company, 84
  system, 83, 92, 93, 96, 98
Environmental
  levels, 134
  matrices, 146
  parameters, 128
Enzymelinked immunosorbent assay
  (ELISA), 180
Enzymes, 57, 65
Epigenetic expressions, 10

Epithelial cells, 107, 115
Escherichia coli, 12, 111, 137, 240
Esters, 112, 128, 134, 139, 146, 179
Ethanol, 107, 111, 115, 174, 184
Ethnicity, 18, 64
Euclidean distance, 235, 237, 238
Exhalation process, 86
Exhaled breath, 6, 7, 61, 69, 82–85, 87–90,
  97–100, 105, 107, 108, 160, 161, 163,
  174, 176–178, 181, 184
  analysis, 82
  condensate (EBC), 108, 168, 169, 171,
    181, 183
  lung cancer, 82
  vapor/condensate (EBV/EBC), 168
Exogenous
  factor, 17
  human volatiles, 9
  source, 5, 12
Exposome, 10
External validation, 197, 212, 215–218
  subset, 215
Extracellular matrix, 59, 60
Extractabrite ion source, 44

F

False discovery rate (FDR), 206
Familywise error rate (FWER), 206
  FWER strategy, 206
Fatty acid
  metabolites, 108
  oxidation, 64
  synthesis, 111
Fecal headspace analysis, 93
Feces analysis, 93
Fetor
  hepaticus, 54
  oris, 54
Field asymmetric ion mobility spectrom-
  etry (FAIMS), 5, 13–15, 175
Field effect transistors (FET), 80
Flavonoids, 83
Flow
  cytometry, 124
  rate, 36, 128, 139, 143
Food

beverage industry, 80
production processes, 141
Fourier transform
infrared spectroscopy (FT-IR), 125, 127,
146–148
near infrared spectroscopy (FT-NIRS),
127, 147, 148
Fourier transforms ion cyclotrons resonance (FT-ICR), 43, 44
Fragmentation
pattern, 45
profiles, 38
FT signal transformation, 45
Function analysis models, 85
Fungal
colony, 112
infections, 112, 113
pneumonia, 113
pulmonary
diseases, 113
interactions, 113
respiratory disorders, 113
species, 113, 125, 126, 150
Fungi, 12, 93, 112, 113, 126, 129, 133,
141, 150, 228, 230, 231, 243
Fungus, 112, 113, 117
cell culture models of volatile production, 112
Furans, 128

G

Gas
analysis, 169
chromatography (GC), 4–6, 14–16, 18,
31–39, 44–46, 54, 108, 113, 115,
125, 134, 144, 146, 147, 151, 169,
170–175, 177, 179–181, 184, 199,
201, 229
GC system, 33, 36, 39
GC-MS analysis, 38, 151, 180, 181
GC-mass spectrometry (GC-MS)
systems, 15, 32, 37, 38, 45, 46,
115, 146, 172, 199
GC-MS data, 15, 45
liquid chromatography mass spectrometry (GC-MS AND LC-MS), 171

mixtures, 32
phase, 32, 39, 170, 172, 175
Gaseous
mixture, 79
phase, 171
Gastric
cancer, 88, 89
intestinal metaplasia (OLGIM), 88, 89
lesions, 88, 199
Gastrointestinal (GI)
illness, 225, 229
tract, 12, 58, 65, 92
Gel electrophoresis, 173
Gene
differences, 114
expression, 67, 114
profiling, 67
Genetic, 13
disorders, 229
mutation, 97
tests, 18
Genome, 12, 13, 116, 198
Geographical
metallurgical samples, 177
origin, 9
regions, 64
Geosmin, 132, 133
Germ cell tumors, 67
Glass impingers, 134, 142, 143
Glucose
consumption, 57
metabolism, 57
Glycolysis, 56, 182, 184
Gold
nanoparticles, 82, 98
standard technique, 15, 117
Graph-clustering, 235, 239, 240

H

Habitual alcohol consumption, 18
Haemophilus spp, 111
Harmonization, 14
Head-and-neck squamous cell carcinoma,
84, 85, 87, 88
Headspace solid phase micro-extraction
(HS-SPME), 146

Health
  conditions, 109, 161, 185, 197
  issues, 126
  status, 13, 105, 106, 110
Healthy
  group, 91
  volatiles, 8
Heart defects, 183
Helicobacter pylori, 54, 65, 161
Helicobacter pylori infection, 54, 161
Helium, 172
Hematology, 18
Hematuria, 18
Hemoglobin, 5
Herbivore infestation, 227
Herbivores, 5, 227
Heterogeneous ionization effect, 40
Hexa-peri-hexabenzocoronene derivatives, 98
Hierarchical
  cluster analysis (HCA), 110
  clustering, 80, 230, 234, 235, 237, 240, 243, 246
  dendrogram, 237
  graph, 239
High altitude pulmonary edema (HAPE), 108
High separation capacity, 146
Histoplasmosis, 113
Homeland security, 80
Hormones, 13
HRV
  infected epithelial cells, 115
  virus, 115
Human
  airway epithelial cells, 114
  B-lymphoblastoid cells, 115
  body sources, 99
  breath, 7, 106, 107, 111–114
  diseases, 20, 225, 228
  genome project, 12
  gut microflora, 92
  health, 4, 124, 225, 229, 241, 243
    care, 228, 241, 243, 246
  leukocyte antigen (HLA), 114, 115
  metabolome database (HMDB), 16
  olfactory displays and interfaces, 203

risk exposure, 126
tissues, 16, 116
volatilome, 6, 8
  databases, 16
Hybrid volatolomics, 99
Hydrocarbons, 111, 139, 140, 144
Hydrogen
  bonding, 34
  cyanide, 111
  peroxide, 181
Hydrophobic side groups, 98
Hydrophobicity, 168
Hypergeometric distribution, 236, 243
Hyphenation, 174
Hypopharynx, 85, 87, 88
Hypothesis testing, 202, 206, 207, 209, 210, 217, 218

I

Idiopathic pulmonary arterial hypertension (IPAH), 108
Immune
  cells/epithelial cell types, 116
  response, 59, 115
  system, 10
Immunochemistry, 124
In vitro, 7, 59, 61, 67, 69, 71, 114, 150, 180
  approach, 94
  clonogenic, 67
  culture cells, 61
  studies, 59–61, 68, 114, 180
In vivo, 7, 60, 61, 63, 67, 69, 71, 116
  imaging assays, 67
  tumor growth/survival assays, 67
Indole, 111
Inductively coupled plasma mass spectrometry (ICP-MS), 176, 177
Infected cells, 113, 115
Infection, 12, 55, 62, 70, 105, 106, 110–113, 115, 116, 162, 169, 178–180, 227, 228
Infectious
  diseases, 178, 229
  non-infectious models, 107
Infiltration, 96

Inflammation, 112, 113, 115
Inflammatory
  bowel disease, 93
  cells, 113
  mediated markers, 105
Influenza
  infection, 179, 180
  studies, 180
Infrared light, 125
Insect pests, 228
Instrumental analysis, 202
Instrumentation, 3, 5, 18, 53, 71, 106, 186,
  199, 200
  analytical techniques, 169
  progress, 106
Intermolecular forces, 34
Internal organs, 89
Invertebrates, 228
Ion
  chromatogram, 45
  fragmentation, 43
  mobility spectrometry (IMS), 5, 54, 92,
    175
  optics, 42, 44
  osculation, 44
  source system, 44
  suppression, 37, 38, 40
  trajectories, 41
  transmission efficiency, 43, 44
Ionic species, 173
Ionization (EI AND CI), 19, 38–40, 44,
  147, 148, 172, 174, 175, 177, 202, 211
  mechanisms, 172
  process, 38
  sources, 148
Irritable bowel syndrome (IBS), 92, 169
Isobutene, 39
Isoprene, 64, 65, 111, 163, 179, 227

K

Ketone bodies, 64
Ketones, 63, 108, 111, 112, 128, 129, 139,
  146, 182
Klebsiella pneumoniae, 138, 240
KNApSAcK, 226, 229–234, 243, 245, 246
  compound ID, 231, 232

  metabolite
    activity database, 233, 234
    ecology database, 231
Krebs cycle, 182
Kyoto Encyclopedia of Genes and
  Genomes (KEGG), 134, 149

L

Lactate production, 57
Laparoscopic radical prostatectomy, 92
Large cell carcinoma, 67
Larynx, 85–88
Leave-One-Out cross validation technique,
  84
Leukemia cells, 57
Libra nose, 82, 84, 91, 94, 98, 100
  system, 84, 94
Linear
  discriminant analysis (LDA), 110, 213
    model, 92
    scoreplots, 213
  variable temperature programming, 37
  velocity, 34, 35
Lipid soluble metabolites, 16
Liquid
  chromatography (LC), 4, 5, 32, 90, 108,
    146, 147, 170, 172–175, 177, 184,
    199, 201–204
  mobile phase, 172
Liver disease, 106, 183
Living organisms, 5, 226
Logistic regression, 92, 183
Low
  dose computed tomography, 84
  limit of detection (LOD), 146
  molecular weights, 125
  thermal mass system (LTMs), 38
Lung
  alveoli, 57, 63
  cancer (LC), 7, 57, 59–63, 66, 82–90,
    96, 97, 167, 169, 177, 181, 182, 184,
    199
  diseases, 65, 82, 105, 110, 183
  health, 110
  lining fluid, 107
  metastasis, 57

tumors, 66, 68, 86
Lymphoblastoid cells, 114, 179
Lymphocytes, 113, 114
Lymphoproliferative tumors, 67

M

Macrophages, 113
Magnesium metabolism disorders, 65
Mahalanobis distances (MD), 95, 225
Malaria, 169, 178–180, 185
Malaria infection, 179
Malignant breast conditions, 89
Mammalian
    cells, 57, 114, 115
        release VOCS, 114
Mass
    accuracy, 43, 44
    analyzer, 19, 31, 32, 40–44, 46, 175
    fragmentation patterns, 38, 40
    measurements, 40, 43
    range applications, 43
    resolution, 42, 43
    spectral libraries, 15
    spectrometry (MS), 5, 6, 14, 15, 17–19,
        36–46, 106, 108–111, 113, 115,
        125, 134, 144, 146, 147, 150, 151,
        169–177, 179–181, 184, 199–204,
        211, 229
        analysis, 43, 45
        fragmentation profiles, 37
        measurement, 40
        spectra, 15, 37
        system, 38–40, 43
    spectrum, 19, 40, 125, 172
    transfer co-efficient, 36
Medical
    metabolomics, 182
    research, 54, 98
Melanoma, 93–95, 199
Melanomas, 94
Menthol/menthone, 174
Metabolic
    changes, 180
    conditions, 228
    differences, 108
    diseases, 229

fingerprints, 200, 210
outcome, 8
pathways, 56, 94, 106, 111, 149, 171,
    185, 243
processes, 93, 176
products, 5, 114
profile, 107, 184, 199
syndrome, 65
Metabolical
    disorders, 65
    pathways, 56
    processes, 63
Metabolite, 5, 7, 10, 12, 13, 16, 17, 19,
    105–110, 114, 115, 125, 133, 149, 160,
    163, 169, 170, 172, 175, 178–182, 184,
    185, 198–200, 203, 204, 206–210, 216,
    226, 227, 229, 231–233, 246
    activity, 229, 231, 233
    analysis, 7
    annotation, 17
    biomarkers, 13
    data, 107
    ecology database, 226, 230–233, 243,
        245, 246
    identifications, 19
    origin develops, 106
    standards, 16
    structures, 16
    volatility, 172
Metabolites identification, 17
Metabolome, 5, 8, 13, 16, 160, 180, 181,
    184, 198, 200, 206
Metabolomics, 3, 5–9, 13–17, 20, 31, 32,
    42, 45, 46, 117, 126, 149, 160, 161, 163,
    169–171, 174, 177–182, 184, 197–201,
    206, 209, 212, 215, 216, 218, 226, 229,
    243, 246
    analysis, 160, 171, 183, 199
    changes, 80, 99
    data, 202
    experiments database, 16
    research, 182
    standard initiative (MSI), 14, 19
    studies, 172
    techniques, 3, 170
Metadata, 3, 12, 15, 17, 211, 218
Metagenomics sequencing, 111

Metaplasia, 88, 89
Metastases, 57, 62
Metastatic
  squamous cell carcinoma, 95
  tumors, 66
Methanol, 107, 111, 174
Methodological problems, 201, 216
Methodology, 8, 53, 59, 60, 68, 151, 202, 214, 216
Methyl thiocyanate, 111
Microbe-human cell interaction, 113
Microbes, 124, 134, 139–142, 150
Microbial, 5, 12, 17, 123–126, 128, 133, 134, 146, 149–151, 228
  communities, 125
  concentration, 123
  fermentation, 125
  growth equals, 133
  identification, 124
  infections, 12
  markers, 150
  origin, 133, 134, 149
  quantity, 124
  species, 124–126, 133, 134
  volatile organic compounds (MVOCs), 17, 123, 125–129, 131–144, 146–151
    concentrations, 128, 133
    data, 141
    measurements, 151
    patterns, 141, 151
    sample collection, 123
Microbials, 9
Microbiome, 5, 9, 112
Microenvironment, 56, 57, 59, 60, 94
Microfluidic
  cartridge, 125
  techniques, 125
Microorganism species, 235, 237, 239, 240, 243
Microscopy, 124, 179
Midstream urine, 91
Migration, 34, 35, 91, 173
Mimic
  non-fungal infections, 113
  tissue and organ systems, 116
Misinterpretation of P-values, 206
Mitochondria, 56, 57

Mobile phase, 33, 35, 36, 172, 173
Molecular
  assays, 67
  chemical techniques, 124
  components, 176
  formula, 233
  orbital, 38
  techniques, 124, 127
  weight, 32, 35, 39, 166, 199, 225, 226, 233
Monoterpenes, 108, 227
Multicenter studies, 53, 65, 67, 70
Multilineage differentiation capacity, 61
Multiple
  breaths, 69
  efficient devices, 185
Multi-sorbent, 144, 145
Multivariate
  analysis, 84, 141, 148, 149, 151, 212
  predictive models, 200
  signal processing, 203
Mutations, 56, 97, 115
Mycobacterium bovis, 178

N

Nanomedicine, 86
NA-NOSE, 82, 85–90, 97, 98, 100
Nanotechnology, 86
Nasopharynx/nasal cavity, 85, 88
Natural product chemistry, 174
Near-infrared spectroscopy (NIR), 125
Neoplasia, 90, 93
Network-based approach, 229
Neutral molecule, 44
Neutralization, 56
Neutrophils, 113, 114
Nitric oxide (NO), 54, 136, 137, 175
Nitrogen compounds, 128
Non-cancerous cells, 12
Non-contaminated site, 128
Non-invasive, 8, 13, 54, 58, 62, 63, 68, 89, 99, 107, 110, 116, 161, 178, 179, 182, 184, 198, 225, 246
Non-malignant
  cells, 56, 57, 60
  diseases, 62, 70

Non-microbial sources, 134
Non-pathogenic bacteria, 238, 239
Non-small-cell lung carcinoma (NSCLC), 97
Non-volatile
  metabolites, 168
  microbial compounds, 146
Normal human diploid fibroblasts(NHDF), 95
Normobaric hypoxia, 108
Nuclear magnetic resonance (NMR), 108, 109, 169–172, 174, 176, 180, 183, 199, 201
  metabolic analysis, 170
  metabolomics, 170
  variables, 183

O

Odor
  fingerprint, 237
  sensing and presentation, 203
Olfactory receptors, 80
Omic technologies, 14
Optical sensors, 80
Oral
  cavity, 65, 85, 87, 88, 162
  gut infections, 106
Orbitap
  mass analyzer, 32, 45
  system design, 44
Ordinary diffusion, 35
Organ/tissue, 81
Organic acids, 128
Oropharynx, 85, 87, 88
Orthogonal projections to latent structures-discriminant analysis (OPLS-DA), 110
Ovarian cancer, 90
Overfitting, 205, 212, 213, 216, 218
  control through validation techniques, 212
Oxidative
  phosphorylation, 57
  stress, 56, 57, 68, 89, 179, 182
Oxidized compounds, 183
Oxygen species, 56

P

Paecilomy cesvariotii, 136
Paenibacillus polymyxa, 137
Parasite, 11, 178
Parasitoids, 227, 228
Partial least squares (PLS), 91, 110, 171, 213, 214
  algorithm, 84
  discriminant analysis (PLS- DA), 91, 110, 213
    algorithm, 82
    model, 91
    scoreplots, 213, 214
Parts per million (ppm), 54, 170, 171, 176, 183
Parts per trillion (ppt), 54, 107, 177
Pathogenic bacteria, 111, 238–240
Pathogenicity, 238, 240, 243
Pathogens, 9, 92, 123, 124, 178, 226–228
Pathophysiologic mechanism, 55
Pentane, 136, 184
Pesticides, 33, 227, 228
P-hacking, 209, 217
Pharmacological manipulation, 17
Pharmacometabolomics, 9
Phenolics, 226
Phenotype, 198
Phenylketonuria, 65
Phosphate metabolism disorders, 65
Phospholipid fatty acids (PLFAs), 124, 127
  PLFAs analysis, 124
Phospholipids analysis, 127
Physical
  examination, 67, 182, 183
  parameters, 80
Physicochemical
  characteristics, 169
  properties, 128
  interaction, 33
Phytopathogens, 228
Plant
  growth, 228, 233, 241
  microbial interactions, 5
  fungus and insect-fungus interactions, 112
Plasma, 9, 18, 99, 161, 177, 182, 198

Plasmodium falciparum, 169, 178, 180
Plasticizers, 134
Plate model theory, 35
Pneumotachograph, 85
Pollinators, 112, 227
Polycyclic aromatic hydrocarbons, 98
Post data acquisition/processing, 45
Post mortem tissues, 18
Postsurgery, 86, 91
Potassium metabolism disorders, 65
Precancerous lesions, 89
Preconcentration, 7, 58, 141
Predictive models, 200, 202, 205, 209, 212, 218
Pre-surgery and post-surgery, 86
Primum non-nocere, 54
Principle component analysis (PCA), 86, 89–91, 95, 96, 110, 149, 171, 203, 204, 212
Probabilistic neural network, 96
Proinflammatory cytokines, 180
Proliferating cells, 59
Proliferation assays, 67
Propylene glycol, 134
Prostate
  biopsy results, 91
  bladder cancer, 91
  cancer (PC), 90
Prostatic cancer, 211
Protein sequences, 16
Proteomics, 13, 160
Proton transfer-reaction mass spectrometry (PTR-MS), 54, 174, 175, 229
Pseudomonas
  aeruginosa, 111, 178, 240
  trivialis, 137
Pulmonary
  arterial hypertension (PAH), 183–185
  arteries, 183
  infections, 113
  nodules, 83, 86
  surfactant system, 180
P-value, 163, 202, 206–210, 217, 237
Pyrolysis, 134

Q

Quadruplor field, 41

Quadrupole
  base systems, 43
  mass filter (QMF), 31, 40–42, 44
    QMF region, 41
  mass selectivity, 45
  time of flight (qTOF), 32, 43
Qualitative/quantitative prediction model, 201
Quality control (QC), 203–205
  process applications, 204
  protocols, 204
  samples, 203–205
Quantitative analysis, 175
Quantum mechanics, 170
Quartz crystal microbalance (QCM), 80–83, 98
Quiescent cells, 59
Q-value, 206

R

Radiation, 5, 9, 57, 67, 68, 146, 175
Radical cation, 38
Radiotherapy, 59, 66, 68
Raman spectroscopy, 176
Randomizing, 211
Reactive oxygen species (ROS), 56, 57
Refectron system, 42
Reproducibility, 70, 95, 96, 106, 127, 147, 148, 163, 172, 201, 204, 216
Respiration collector for in vitro analysis (ReCIVA), 166–168
Respiratory
  diseases, 6, 123, 124, 177, 184, 185
  distress, 70, 180
  function, 58
  illnesses, 182
  tract, 180
Response Evaluation Criteria in Solid Tumors (RECIST), 85
Resultant model, 212
Retention time (RT), 19, 39, 45, 46, 172, 173
Retrospective longitudinal repository studies, 56
RF mass filters, 43
Rhizobacteria, 228

Rhizopus stolonifer, 138
Routine spirometry, 86, 177

S

Saliva, 9, 62, 82, 99, 161, 167, 168, 177, 182, 198
Salivary
  gland tumor, 67
  glands, 87
Sample
  collection, 69, 71, 128, 161, 211
  concentration, 35
  derivatization, 172
  introduction, 18
  matrix, 33, 40, 169
  preparation, 142
  storage, 69, 201
Samplig
  collection, 151
  extraction technique, 18
  method, 53, 69, 128, 161, 163 169
  methodologies, 125
  process, 18, 160, 162
  techniques, 3, 123, 126, 163, 186
Sarcomatoid carcinoma, 67
Scedosporium apiospermum, 133
Screening
  programs, 66
  technique, 66
Secondary electrospray ionization-mass
  spectrometry (SESI-MS), 174
Secretory cells, 92
Selected ion flow tube mass spectrometry
  (SIFT-MS), 54, 174, 175, 184, 229
Selective
  ion monitoring (SIM), 41
  reaction monitoring (SRM), 41
Self-organization mapping (SOM), 110
Semiochemicals, 225, 227, 228
Sensor array, 86
Serum, 9, 18, 58, 95–97, 99, 161, 180, 198, 201, 211
Sesquiterpenes, 112, 134
Signal processing, 199, 200, 205
Single breath canister, 163
Sinusoidal signals, 45

Skin headspace analysis, 93
Skin surface, 93, 94
Sleep apnea, 183
Sludge dewatering site, 132
Small cell carcinoma, 67
Smoking habit, 5, 12, 18
Snoring operation, 88
Sodium metabolism disorders, 65
Solanum lycopersicum, 232
Solid phase microextraction (SPME), 18, 115, 146, 147, 168, 169, 172, 176, 177, 229
Solvent extraction, 139, 140, 146
Species-species interaction, 229
Specific antigen testing, 92
Spectconnect, 45
Spectral
  deconvolution, 19
  intensity bioaerosol sensor (SIBS), 125
Spectrometry, 5, 8, 39, 40, 43, 106, 111, 127, 144, 169, 175, 180, 199, 211, 229
Spectroscopic techniques, 176
Spectroscopy, 125, 127, 148, 169, 170, 173, 175, 176, 199
SpinCon sampler, 143
SpiroNose system, 85
Sputum, 82, 99, 110, 182
Squamous cell carcinoma, 11, 67
Staphylococcus aureus, 111, 137, 178, 240
Stationary phase, 32–36, 38, 172
  composition, 33, 34
Statistical
  analysis, 109, 148, 150, 170, 171, 217, 237
  methodological pitfalls, 197
  models, 110
  techniques, 110
  tools, 109, 110
Stearothermophillus, 137
Stem cell properties, 61
Stenotrophamonas spp, 111
Steroids, 33
Streptococcus pneumonia, 133
Streptomyces citreus, 138
Structure definition file (SDF), 236, 242
Sulphur compounds, 128
Support vector machine (SVM), 87, 110

Surface acoustic wave (SAW), 80, 83
Systemic acquired resistance (SAR), 227

T

Tabular file formats, 19
Tanimoto coefficient, 230, 236, 242, 243
Targeted therapies, 59, 66, 68, 115
Terpenes, 128, 134
Terpenoids, 111, 112, 226
Therapeutic
  intervention, 13
  options, 68
Thermal
  desorption (TD) tubes, 131, 134, 143,
    144, 146, 151
  stability, 168
Time of flight (TOF), 31, 42–44, 181, 211
Tissue
  cells, 12
  samples, 62, 182, 184
Tobacco components, 134
Toluene, 115, 174
Tonsillectomy, 88
Transcriptome, 13
Transcriptomics, 13, 160
Transduction features, 83
Transurethral resection, 92
Treatment, 13, 55, 60, 66–68, 85, 90, 91,
  132, 177–179
Tricarboxylic acid (TCA) cycle, 184
Trimethylamine, 115
Triple quadrupoles (QQQ), 31, 41
Tritrophic interactions, 227
Tumor, 12, 56, 57, 59–68, 83, 85, 88–90,
  94–96
  cells, 12, 57, 59, 64, 96
  recurrence, 62, 63
  suppressor genes, 56
Tumoral characteristics, 59
Types of cancer, 58, 62–65, 67, 95

U

Ultrafast gas phase, 38
Ultra-high resolving power, 43, 45
Ultrasound, 184

Ultraviolet–visible spectroscopy (UV-vis),
  173
Univariate hypothesis testing, 200
Upper respiratory infections (URIs), 105
Urban and agricultural environments, 124
Urea breath test (UBT), 54
Urinary system, 63
Urine
  chemistry, 18
  headspace analysis, 92
  midstream, 91
  samples, 91, 92, 204
  volatiles analysis, 91
Urological
  malignancies, 199
  pathologies, 91

V

Vaccinated group, 180
Vaginal fluid, 9
Van Deemeter equation, 31, 35, 46
Vapor
  pressure, 225, 226
  sensing instrument, 98
Velocity, 36
Ventilation, 70
  perfusion scan, 184
Viral, 105, 113, 115, 117, 179, 180
  analysis, 54
  classification, 240
  data, 230, 231, 237
    base development, 226, 229
  emitting species, 234
  infections, 105, 115, 179
  non-VOC metabolites, 106, 107
  respiratory infection changes (VOCs),
    113
  sample analysis, 181
  signature, 80, 111, 113, 179
Volatile, 9, 10, 33, 55, 57, 58, 60, 62,
  65, 68, 91–94, 96, 98, 112, 115, 166,
  227–230
  compounds, 15, 35, 69, 94, 113, 146,
    147, 170, 172, 174, 198
  emission, 225
  fraction, 13

organic compounds (VOCs), 3–10,
    15–17, 31, 32, 39, 46, 53–65, 68–71,
    80, 82, 93, 96–99, 105, 107, 110–117,
    125, 126, 133, 134, 141, 146, 149,
    151, 161, 162, 165–169, 174–179,
    198, 225–231, 233, 235–238,
    240–243, 246
    analysis, 198
    emitting species, 232
    pattern, 59
    transformation, 65
    profile origin, 64
profiles, 228
stereoisomers, 174
Volatilome, 5, 6, 8, 15–17, 20
    volatolome, 5
Volatilomics, 3, 7, 10, 13, 14, 17, 19

W

Warwick olfaction system wastewater
    treatment plants (WWTP), 124, 128,
    131, 132, 149

Water
    metabolite content, 107
    pharmaceutical preparations, 148
Weather
    conditions, 128, 141, 150
    station, 141
Wideband Integrated Bioaerosol Sensor
    (WIBS), 125
Wind
    direction, 141, 150
    speed, 141, 150
World Health Organization (WHO), 4, 55,
    66, 67, 185, 202

X

Xenobiotic, 4
Xylene, 174

Z

Zones, 36